II International Congress of Mechanical Engineering and Agricultural Science (CIIMCA 2015)

IOP Conference Series: Materials Science and Engineering Volume 138

Floridablanca, Columbia
7 – 9 October 2015

ISBN: 978-1-5108-2674-8
ISSN: 1757-8981

Printed from e-media with permission by:

Curran Associates, Inc.
57 Morehouse Lane
Red Hook, NY 12571

Some format issues inherent in the e-media version may also appear in this print version.

Printed by Curran Associates, Inc. (2016)

For permission requests, please contact the Institute of Physics
at the address below.

Institute of Physics
Dirac House, Temple Back
Bristol BS1 6BE UK

Phone: 44 1 17 929 7481
Fax: 44 1 17 920 0979

techtracking@iop.org

Additional copies of this publication are available from:

Curran Associates, Inc.
57 Morehouse Lane
Red Hook, NY 12571 USA
Phone: 845-758-0400
Fax: 845-758-2633
Email: curran@proceedings.com
Web: www.proceedings.com

TABLE OF CONTENTS

IOP Conf. Series: Materials Science and Engineering **138** (2016) 011001 doi:10.1088/1757-899X/138/1/011001

Editorial

The 2nd International congress of Mechanical Engineering and Agricultural Science (CIIMCA 2015) looked to exchange new scientific knowledge on mechanical engineering, agricultural science, materials science and related topics. The CIIMCA 2015 was organized by the Mechanical Engineering Faculty of the ´Universidad Pontificia Bolivariana´ of Bucaramanga. It was held from 7th to 9th October 2015 at ´Universidad Pontificia Bolivariana´ of Bucaramanga, Colombia. The meeting aimed to provide a high level international conference for researchers in mechanical engineering and related themes.

The CIIMCA brought together the latest results and discussed new advances over a broad range of topics dealing with vibration modeling & testing, renewable energy and thermofluids analytical, science and drying technology, materials science, automation and agricultural science and engineering education as well as case studies. We did our best to provide the delegates with an environment conducive to the exchange of knowledge, networking and making new friends.

We want to thank our colleagues who contributed considerably to the event by organizing topical symposia and by helping us to launch this important meeting. We are truly grateful for their efforts.

PhD. Aduljay Remolina-Millán.
Chairman, International Congress of Mechanical Engineering and Agricultural Science Congress –CIIMCA 2015.

PhD. Rolando Guzmán-López.
Chairman, International Congress of Mechanical Engineering and Agricultural Science Congress –CIIMCA 2015.

IOP Conf. Series: Materials Science and Engineering **138** (2016) 011001 doi:10.1088/1757-899X/138/1/011001

Organised by

Mechanical Engineering Faculty
Universidad Pontificia Bolivariana of Bucaramanga.
Floridablanca, Colombia.

Conference Organisation

Chairmen

PhD. Aduljay Remolina-Millán.
Associate professor
Mechanical Engiennering Faculty
Universidad Pontificia Bolivariana of Bucaramanga

PhD. Rolando Guzmán-López.
Mechanical Engiennering Faculty
Universidad Pontificia Bolivariana of Bucaramanga

Acknowledgments

The organizing committee wants to thank the Editorial Committee for their support, participation and valuable contribution to CIIMCA 2015.

The organizing committee wants to thank the Mechanical Engineering Faculty and the "Universidad Pontificia Bolivariana seccional Bucaramanga" for its support and funds given to the development of CIIMCA 2015.

We want to thank our sponsors and contributors for their support and cooperation in the organization of the CIIMCA 2015 Congress.

Finally, the organizing committee wants to thank all attendees of CIIMCA 2015, thanks for your enthusiastic participation.

The organizing committee wants to specially thank Sarah Toms, Conference Commissioning Editor of the IOP Conference Series for her support and help in the process of publishing the papers from CIIMCA 2015 in the IOP Conference Series.

IOP Conf. Series: Materials Science and Engineering **138** (2016) 011002 doi:10.1088/1757-899X/138/1/011002

Peer review statement

All papers published in this volume of *IOP Conference Series: Materials Science and Engineering* have been peer reviewed through processes administered by the proceedings Editors. Reviews were conducted by expert referees to the professional and scientific standards expected of a proceedings journal published by IOP Publishing.

IOP Conf. Series: Materials Science and Engineering **138** (2016) 012001 doi:10.1088/1757-899X/138/1/012001

Preparation of potato starch microfibers obtained by electro wet spinning

W Cárdenas[1,4], E Y Gómez-Pachon[2,4], E Muñoz[1,4], and R Vera-Graziano[3]

[1] Escuela de Ciencias Químicas, Facultad de Ciencias, Universidad Pedagógica y Tecnológica de Colombia-UPTC. Avenida Central del Norte 39-115 Sede Central Tunja–Boyacá–Colombia

[2] Escuela de Diseño Industrial, Facultad sede Duitama, Universidad Pedagógica y Tecnológica de Colombia-UPTC. Carrera18 con Calle 22 Duitama-Boyacá-Colombia

[3] Instituto de investigaciones en materiales-IIM, Universidad Nacional Autónoma de México-UNAM. Ciudad Universitaria, circuito exterior s/n 04510 México D.F-México

[4] Grupo de Investigación en desarrollo y aplicación de nuevos materiales-DANUM, Universidad Pedagógica y Tecnológica de Colombia-UPTC.

Email: edwin.gomez02@uptc.edu.co

Abstract. Starch is one of the most abundant biopolymer in nature. It has been primarily used as a thickener in the food industry. Starch is found in greater amounts in the potato tubers, which is one of the largest food productions in the region of Boyacá-Colombia. Thus, potatoes are a viable source of starch. The main objective of this study is the preparation and characterization of native starch´s microfiber by electro wet-spinning technique. The parameters that were changed for each treatment were as follows: the amount of potential applied to the solution, the distance between the needle and the collector and the rate of injection of the solution in order to determine the physical and chemical properties of the membranes, conformed by potatoes starch microfiber. Diverse instrumental analysis techniques were applied. They were: Scanning Electron Microscopy (SEM) to determine the morphologies and diameters of microfibers, Fourier Transform Infrared Spectroscopy (FTIR) to determine the chemical changes, Thermogravimetric Analysis (TGA) and Differential Calorimetry Scanning (DSC) to obtain the thermal transitions and the temperatures of useful. The microfibers were analysed in order to determine their structural properties and thus define the range of application. In conclusion, potatoes starch microfibers were obtained with average diameters of 15, 17, 23 and 25 micrometres, besides the fibers presented a degradation temperature of 304 °C, indicating that fibers are available with diameters of small scale, with good thermal properties. This study will enable the implementation of the microfibers to obtain bio packaging for food products and other applications.

1. Introduction

Polymeric Nano and microfibers have been studied enough due to their characteristics which present high porosity, mechanical strength, flexibility and adhesion to other materials and high surface energy due to tiny diameters. Some other outstanding microfibers polymers' properties such as the mechanical, physical and chemical have been reported in several papers. These exalted properties are important for the development of new materials for industrial and technological applications. One of the most used techniques for preparation in nano-microfibers is the electrospinning. This is due to the versatility to prepare continuous fibers in diverse sizes: from nano to the micro size [1, 2].

The main parameters of electrospinning are power supply of high voltage, rate of injection of the solution and distance between needle and metallic collector. The power supply of high voltage charges the polymer solution accelerating it from the needle to a collector. A jet of solution is ejected toward the metal plate due to the electric field force which is greater than the surface tension of the polymer solution. The jet of the solution is exposed to the atmosphere over the course of needle to collector, allowing the solvent to evaporate. This is to obtain solid polymer fibers with small diameters in the collector [3-5]. This technique in which fibers with small-scale diameters are formed by use of electrostatic forces, was discovered many years ago[6-8], but in the last decade research in this area has been increasing given that many studies for the production of nonwoven webs from a variety of synthetic polymers to be used in different applications such as filters, textiles, tissue scaffolds, drug delivery, sensors and nanocomposites[9-14].

Also electrospun from natural polymers such as starch and cellulose have been studied, being that these polysaccharides have rapid degradation, high biocompatibility and are not harmful to the environment. These natural polymers have been studied for the manufacture of products that replace fossil fuel-based materials [15, 16]. Starch is a major natural polymer, besides being abundant in nature, is also a great industrial utility either in the food, pharmaceutical, textile, paper, among others [17-20]. One of the main advantages for natural polymers is its easy access because there are various sources that have concentrations of more than 70% of polysaccharide. Among the most food used for the extraction of starch are corn, potatoes, barley and wheat [21-24].

In previous studies Kong, L et al [25] conducted electrospinning of starch from different sources such corn, potatoes, beans, rice and barley. They considered several grades of concentration of amylose and amylopectin, composites in which variation of concentration in the starch, offers different qualities as high heat resistance as described by Barichello et al [25]. Also Kong et al studied parameters of electro wet-spinning for obtaining corn starch fibers with diameters around 1 a 5 micrometers. Besides Kong, L et al [25-27] obtained starch microfibers the electrospinning equipment was configured to electro wet-spinning where the plate collector was immersed in pure ethanol with the final purpose of retirement of dimethyl sulfoxide dissolvent. However, they used industrial starch for electrospun microfibers and they have not been made electro wet-spinning from nature starch from the potatoes.

For this reason, the proposal of this paper is to obtain the starch potatoes microfibers with the technical of electro wet-spinning from the natural potatoes and special from the region of Boyacá like the variety Diacol Capiro (R12) which its high amylose gives the starch high thermal resistance, this being quite important for the preparation of new biodegradable materials[26] with the objective for industrial applications.

2. Experimental

2.1. Materials

The extraction of starch was carried out from to the potato variety Diacol Capiro, also known as R-12, cultivated in Boyacá-Colombia. The solvent for the preparation of starch solution was dimethyl sulfoxide (DMSO, CAS: 67-68-5), being acquired from the marketing firm Merck at a concentration of 99%. Also ethanol (CAS: 64-17-5) was used. It was purchased from Sigma-Aldrich trading house at a concentration of 99.8%. The extraction of the starch is carried out following the methodology proposed by Vasanthan et al and Aristizabal et al [1, 6]. The solution obtained is allowed to settle for 12 hours and the precipitate was washed with deionized water and subjected to a drying air flow at 40 °C for 8 hours.

2.2. Electro wet-spinning

The equipment used for this work was the electrospinning which was configured to wet. It was constructed at Universidad Pedagógica y Tecnológica de Colombia- hereafter UPTC, (see Figure 1). The power supply used is HV 350R provided by the company Unlimited Company Amherst, the injector was provided by the company Syringe Pump Company Pump Systems Inc model NE model 4000 injector. The collector plate was immersed in a solution of ethanol-water 70:30, because this solution dissolves the DMSO and so the microfibers collected are starch with high purity. The starch solution used in electro wet-spinning process was prepared according to Kong et al [5], with a starch concentration of 10% w/w, due to starches with a percentage higher than 70% amylose are completely dissolved in DMSO in a range of 10-20% w/w of starch.

Figure 1. Electro wet-spinning equipment belonging to UPTC.

2.3. Techniques used for the characterization

The fibers were analysed by infrared Fourier transform spectroscopy with attenuated total reflectance (ATR-FTIR). The equipment used for the analysis was the ATR-Thermo Scientific Nicolet iS5, making a total of 32 scans in a range from 4000 to 400 cm-1.For the Thermal Analysis was used the DSC and TGA about the starch potatoes fibers. This technique was performed in a calorimeter SDT Q600 TGA / DSC TA Instruments. The analysis was performed with a heating rate 10 °C/min, 18 to 400 °C. For the morphological determination and for the porosity of the fibers was performed a technique with scanning electron microscopy-SEM. These micrographs were made through JEOL-JSM-7600F (Japan) equipment Tungsten filament with a voltage difference of 20.0 KV. Photographs taken for analysis of the fibers contained increases from 60x to 650x. To carry out the estimation of fiber´s diameters, the micrographs were analysed using the Image J software.

3. Results

3.1. Infrared Spectroscopy

The fibers were analysed by ATR-FTIR to observe if the electro wet-spinning process causes chemical changes in the fibers. The native starch potatoes of powder were taken as reference to the infrared spectrum and so, used to determine whether there are changes in the spectrum of the fibers. In Figure 2, the spectra obtained fibers starch and native starch powder are presented. The IR spectrum for fibers of native starch presented one band at 3289 cm−1 related to the OH group bending [27] and at 1148,60; 1076,68 and 1000 cm−1 corresponding to –C–O–C– stretching vibration of glucose [28]. The bands found between 940 and 575 cm−1 were attributed to the whole glucose ring stretching vibrations. The band at 1635,26 cm−1 is due to tightly bound water present in the starch [29].The signal at 2926,59 cm−1 was the characteristic vibration of C–H stretches [30].

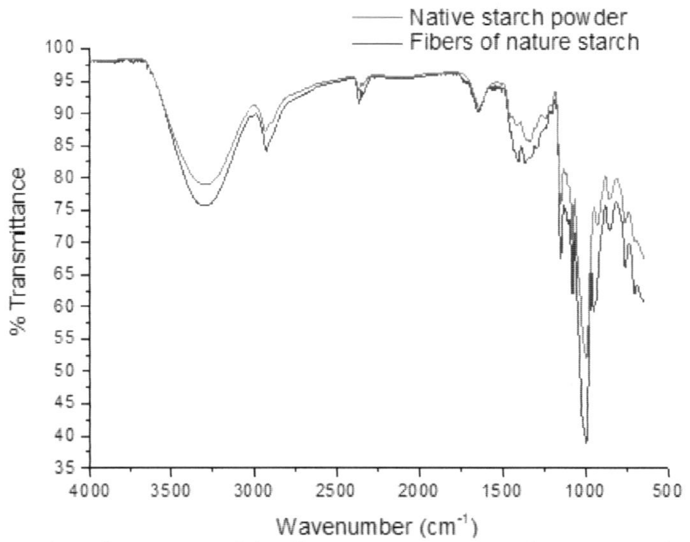

Figure 2. Infrared spectrum of the fibers potato starch and potato starch powder

For the of native starch, the IR spectrum presents the band of the OH group bending at 3296,59 cm-1. The C-O-C stretching vibration of glucose presents signals at 1148,15; 1077,83 and 1000 cm-1 and the bands attributes to the glucose ring stretching vibrations are between 937-580 cm-1. At 1652,15 cm-1 is the bound water present in the starch and the band characteristic vibration of C-H stretches is found at 2921,93 cm-1. As shown in Figure 2, the electrospun fibers of native starch do not present chemical changes to the native starch powder. The bands found in both spectra are described in Table 1. Given the similarity of the signals obtained, we can confirm that the starch does not undergo chemical changes as a result of electro wet-spinning.

Table 1. Signals encountered in the infrared spectrum of the fibers of native starch and starch.

Starch Fibers (cm^{-1})	Native Starch (cm^{-1})	Type of vibration
3289,08	3296,59	Stretch wáter
2926,59	2921,93	C-H stretching
1635,26	1652,15	O-H bending
1148,60; 1076,68 y 1000	1148,15; 1077,83 y 1000	Stretch C-O-C

3.2. Thermal Analysis- TGA

To observe the thermal properties belonging to the starch potatoes fibers an analysis of TGA was conducted and the results are shown in Figure 3.

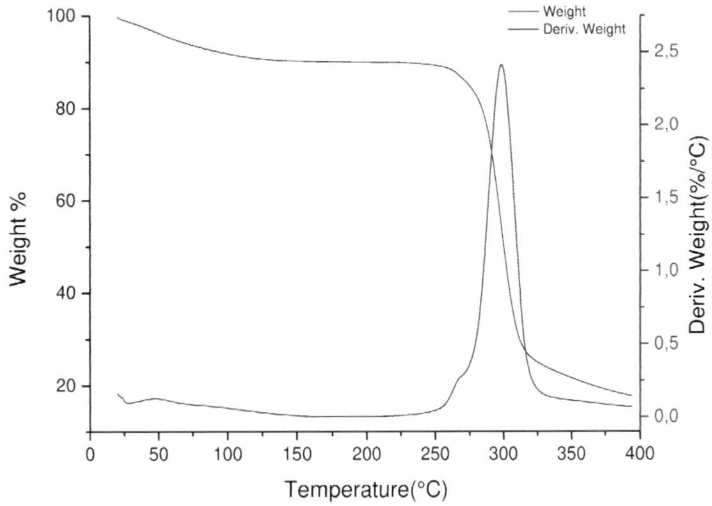

Figure 3. TGA diagram of starch potatoes fibers

In thermogravimetric analysis, the loss in mass due to volatilization of the degradation products is monitored as a function of temperature. It is possible to observe the loss of humidity of starch potatoes microfibers at 100°C corresponding to 11% of loss of total weight. After evaporation, the weight of the fibers stay constant until 250°C, where it is observed the beginning of loss of the weight of the fibers that correspond to the beginning of decomposition temperature of the starch fibers. The blue line observed in the figure, correspond to the derivative of weight in function of temperature. Here the temperature of fast degradation of fibers at 304°C can be seen. The fast loss of weight occurred at 315°C, in this transition point the fibers have lost near of 60% as compared with the initial weight. This analysis indicates the maxim temperature for usefulness (250°C) that must be considered for the design of elements with this material.

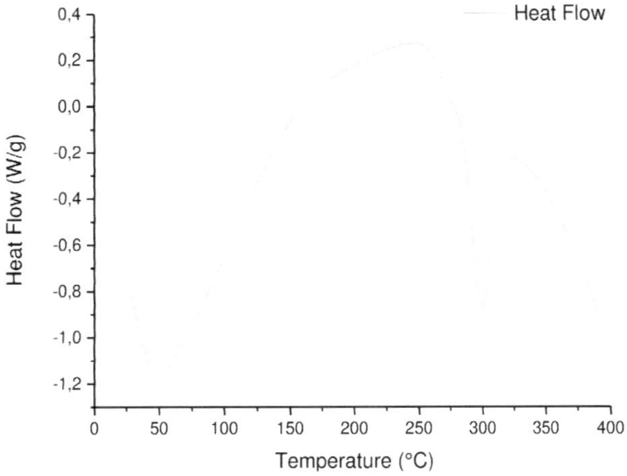

Figure 4. DSC of starch potatoes fibers.

In figure 4 it is possible to observe the energy absorptions that presents the native starch potatoes fibers with the increase of the temperature. Here the gelatinization temperatures of starch potatoes microfibers can be seen which are endothermic signals. The first signal appeared at 50°C, where the first stage of absorption of enthalpy energy of 81.16J/g is presented. This energy is absorbed by the fibers of starch for the dissociation of amylose (amorphous part) present in the starch. The second endothermic peak appeared at 303°C with enthalpy energy of 120 J/g. It occurs when this energy break the amylopectin bound. When the process increases the temperature, the fibers absorb the heat and generate the dissociation of amorphous phase (amylase). When the temperature is increasing, this generates the melting of a second component of starch potatoes, known, in this case, as amylopectin. The easy dissolution of amylase starch potatoes fibers at a low temperature happen because the hydrogen bonds are not that strong, while the bond of amylopectin is most stable, therefore it required higher temperature. Figure 4 confirms what is reported in several papers. It is to say that starch potatoes Diacol capiro (R-12) presents a high temperature resistance [26], so it is a good candidate for the fabrication of biodegradable material and industrial applications that require temperature above 100 C°.

3.3. Morphological analysis

The Morphological analysis of the fibers was carried out to four treatments of fibers. The Electro wet-spinning parameters used are shown in Table 2. In Figure 5 it is showed the micrograph taken to 4 treatments mentioned.

Table 2.Parameters used in the 4 treatments.

Parameters	Treatment 1	Treatment 2	Treatment 3	Treatment 4
Voltage (Kv)	10	10	8	8
Rate injection (ml/h)	2	2	2	4
Distance needle-collector (cm)	8	6	6	6

In figure 5 it can be observed that all fibers obtained present roughness, cross-linking and agglomerations in some areas. This can indicate the effects of electro wet-spinning process because despite the variety of possible conditions, the contact of the jet of solution of starch with the ethanol, in the moment of deposition in the collector, does not generate completely the solidification and then the fibers were still wet and it creates the cross-link between the fibers. On the other hand, the agglomerations are evidence of the no continuity in the process. Therefore some polymers leave the needles with major o less volume.

In the first treatment, it is possible to denote that the fibers are fused. This could be due to the presence of solvent in the time of deposition that was not evaporated completely in the process of spinning, as consequence of high voltage. In the fibers of second treatment, it can be observed roughness, inhomogeneous in the size of the diameter and defined fibers. In contrast with the first treatment, here the process had less time and therefore, the fibers are thicker (25 micrometers) than the first treatment (17 micrometers).

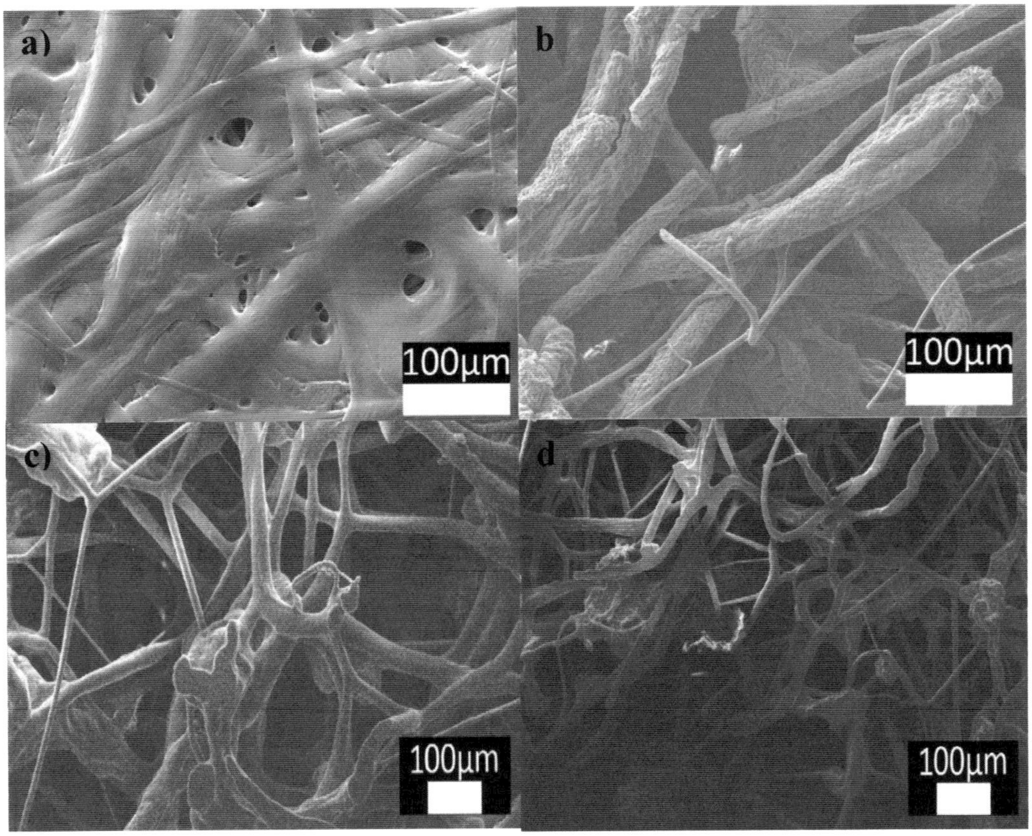

Figure 5. Micrograph of SEM starch potatoes fibers taken at the Instituto de Investigaciones en Materiales IIM of UNAM, a) treatment 1 (250X), b) treatment 2 (250X), c) treatment 3 (200X), d) treatment 4 (100X).

The fibers of the third treatment presented similar characteristic to second treatment such as continuous and defined fibers. Also it is possible to observe that the diameters of fibers in this treatment are more homogeneous, interlinking and smaller (15 micrometers) in comparison with the second treatment. This is explained because of the decrease of the voltage that allowed the jet of the solution to leave the capillary with less force and more time in the spinning. The fibers of the fourth treatment are similar to the third in the homogeneity in the diameters, the interlinking and they presented an increase in the diameters (25 micrometers). This is possible due to the increase in the rate of injection.

In studies carried out by Kong et al [31] with industrial corn starch, they obtained fibers with diameters between 1 and 5 micrometers, with a smooth morphology and homogeneous diameters. The difference with our results in the inhomogeneous morphology is due to the starch potatoes variety, Diacol Capiro (R12), presents different concentrations of amylose and amylopectin in comparison with starch corn. In consequence, it is generated a change in the viscosity of polymeric solution. In addition, these two components (amylose and amylopectin) influence in the starch thermal properties which was showed in the TGA. Another reason for this difference in diameter is due to the fact that our ethanol had a percentage of water that caused a time of solidification of fibers different.

The treatment with more homogeneity was treatment 3 and although the studies of Milasius et al [32] obtained fibers (125 nanometers) thinner in the order of nanometers, the fibers in our work were obtained from natural starch potatoes with voltage in the order of 6-8 Kv, while Milasius et al used voltage in the order of 75 Kv. They also used polyvinyl alcohol such dissolvent with addition with ethanol and collected the fibers with drum.

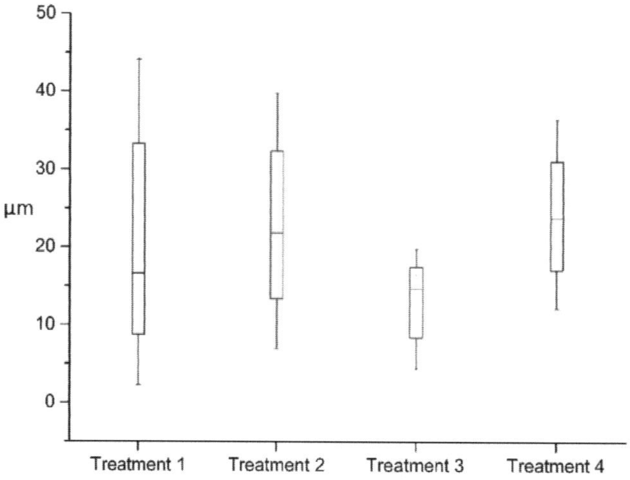

Figure 6. Boxplot about 4 treatments.

The analyses of diameter of starch potatoes fibers was measure by software ImageJ and the results are shown in figure 6 with the boxplot, where the upper diameters, the average and the lower diameter of the fibers in each treatment are represented. In figure 6 it can be observed that the third treatment with a voltage of 8 Kv, distance between needle and collector of 6 cm and rate of injection of 2 ml/h, presented the lowest diameter average (15 micrometer) and lowest variation (among 5 y 20 micrometers) in comparison with other treatments.

The treatment 1showed a mean diameter of 17 micrometer, similar to the treatment 3, but showed a more variation (3 - 44 micrometers). This difference in the variation may be due to the fact that the first treatment used higher voltage (10 Kv) with distance 8 cm generate more mobility of the fibers in the path between needle and collector and as a consequence the fibers are deposited in several places of the collector and it generates the variation of diameters of fibers.
The treatment 4 (8 Kv, 6 cm and 4 ml/h) generates an average diameter of 25 micrometers that exceeds the others treatments. Here the only difference in the treatment is in the increase of rate of injection of 2 ml/h, which demonstrates the influence of the rate injection about the diameter of starch potatoes fibers. Such influence may explain that higher rate injection causes major volume of the solution in the needle, therefore the force generated by the field electric does not decrease diameter of jet generating an increase in the diameter of the fibers deposited in the collector. However, in this case the variation of diameters is lower in comparison with the treatment 2, this is due to more voltage that makes the polymer accelerates toward to collector. Then, the fibers are deposited faster and dispersed without enough time for stretching.

4. Conclusions

This study proves that native potato's fibers from the Region of Boyacá Colombia, with diameters in the size of micrometers can be obtained. It also shows the effect of the three parameters used in the electro wet spinning, being the parameters of the third treatment the one that showed homogeneous continuous fibers, and high interlinking. It also notes that the deposition and washing with ethanol for the separation of solvent of the fibers, in the electrospinning process, was fundamental to obtain fibers. It was also noted that after the electrospinning, the fibers do not showed a change in the chemical structure as evidenced with the FTIR, Besides that, the test of DSC and TGA evidence that the fibers show a high resistance to heat (304°C is the maxim temperature of degradation), which is of quite important for the use of this material in the industry. Because the maxim temperature for useful found for the starch potatoes microfibers was 250°C that must be considered for design of elements with this material

5. Acknowledgments

The present study was carried out under the cooperation agreement between UPTC - IIM-UNAM and it was supported by the projects PAPIIT-UNAM/IN108930, CONACYT/ CNPq174247 in México and the projects at UPTC with SGI codes: 1573 y 1769 in the UPTC-Colombia.

References

[1] Huang Z-M et al 2003 A review on polymer nanofibers by electrospinning and their applications in nanocomposites *Composites Science and Technology*. **63** 2223-2253
[2] Chronakis IS 2015 Micro- and Nano-fibers by Electrospinning Technology: Processing, Properties, and Applications, in Micromanufacturing Engineering and Technology (Second Edition), Y Qin, Editor., William Andrew Publishing: Boston **22** 513-548
[3] Kim IG et al 2015 A comprehensive electric field analysis of cylinder-type multi-nozzle electrospinning system for mass production of nanofibers *Journal of Industrial and Engineering Chemistry*. **31** 251-256
[4] Deitzel JM et al 2001Controlled deposition of electrospun poly(ethylene oxide) fibers *Polymer*. **42** 8163-8170
[5] Fong H, I Chun and DH Reneker 1999 Beaded nanofibers formed during electrospinning. *Polymer*.**40** 4585-4592
[6] Morton WJ 1902 Method of dispersing fluids, *Google Patents*
[7] Cooley J 2006 *Improved methods of and apparatus for electrically separating the relatively volatile liquid component from the component of relatively fixed substances of composite fluids.* United Kingdom Patent, 1900. **6385** 19
[8] Cooley JF, *Apparatus for electrically dispersing fluids*. 1902, Google Patents.
[9] Pant, HR et al 2011 Photocatalytic and antibacterial properties of a TiO2/nylon-6 electrospun nanocomposite mat containing silver nanoparticles *Journal of Hazardous Materials* **189** 465-471
[10] Pant HR et al 2013 Synthesis, characterization, and mineralization of polyamide-6/calcium lactate composite nanofibers for bone tissue engineering. *Colloids and Surfaces B: Biointerfaces*. **102** 152-157
[11] Burger, C, BS Hsiao, and B Chu 2006 nanofibrous materials and their applications. *Annual Review of Materials Research*. **36** 333-368
[12] Neppalli R et al 2013 The effect of clay and of electrospinning on the polymorphism, structure and morphology of poly(vinylidene fluoride) *European Polymer Journal* **49** 90-99
[13] Neppalli R et al 2010 Poly(ε-caprolactone) filled with electrospun nylon fibres: A model for a facile composite fabrication *European Polymer Journal* **46** 968-976

IOP Conf. Series: Materials Science and Engineering **138** (2016) 012001 doi:10.1088/1757-899X/138/1/012001

[14] Unnithan AR et al 2012 Wound-dressing materials with antibacterial activity from electrospun polyurethane-dextran nanofiber mats containing ciprofloxacin HCl *Carbohydrate Polymers* **90** 1786-1793

[15] Silva RM et al 2016 Dental glass ionomer cement reinforced by cellulose microfibers and cellulose nanocrystals *Materials Science and Engineering: C* **58** 389-395

[16] Sonia A and K Priya Dasan 2013 Celluloses microfibers (CMF)/poly (ethylene-co-vinyl acetate) (EVA) composites for food packaging applications: A study based on barrier and biodegradation behavior *Journal of Food Engineering* **118** 78-89

[17] Rodrigues, A and M Emeje 2012 Recent applications of starch derivatives in nanodrug delivery *Carbohydrate Polymers* **87** 987-994

[18] Torres FG et al 2011 Biocompatibilty of starch-based films from starch of Andean crops for biomedical applications *Materials Science and Engineering: C* **31** 1737-1740

[19] Hung PV, NL Vien and NT Lan Phi 2016 Resistant starch improvement of rice starches under a combination of acid and heat-moisture treatments *Food Chemistry* **191** 67-73

[20] Buddrick O et al 2015 The effect of fermentation and addition of vegetable oil on resistant starch formation in wholegrain breads *Food Chemistry* **180** 181-185

[21] Moreno O L Atarés and A Chiralt 2015 Effect of the incorporation of antimicrobial/antioxidant proteins on the properties of potato starch films *Carbohydrate Polymers* **133** 353-364

[22] Villamonte G, V Jury, and M de Lamballerie 2016 Stabilizing emulsions using high-pressure-treated corn starch *Food Hydrocolloids* **52** 581-589

[23] Yangcheng H et al 2015 Pysicochemical properties of tibetan hull-less barley starch *Carbohydrate Polymers* **137** 10

[24] Rosicka - Kaczmarek J et al 2016 Composition and thermodynamic properties of starches from facultative wheat varieties *Food Hydrocolloids* **54 Part A** 66-76

[25] Barichello V et al 1990 Low Temperature Sweetening in Susceptible and Resistant Potatoes: Starch Structure and Composition *Journal of Food Science* **55** 1054-1059

[26] Pardo Cuervo, OH, JC Castañeda Mancipe and CA Ortiz 2014 Caracterización estructural y térmica de almidones provenientes de diferentes variedades de papa. **62** 7

[27] Fang JM et al 2002 The preparation and characterisation of a series of chemically modified potato starches *Carbohydrate Polymers* **47** 245-252

[28] Fang JM et al 2004 The chemical modification of a range of starches under aqueous reaction conditions *Carbohydrate Polymers* **55** 283-289

[29] Kačuráková M and RH Wilson 2001 Developments in mid-infrared FT-IR spectroscopy of selected carbohydrates *Carbohydrate Polymers* **44** 291-303

[30] Chi H et al 2007 Synthesis of dodecenyl succinic anhydride (DDSA) corn starch Food *Research International* **40** 232-238

[31] Kong L and GR Ziegler 2014 Fabrication of pure starch fibers by electrospinning *Food Hydrocolloids* **36** 20-25

[32] Milašius R 2010 Investigation of the Possibility of Forming Nanofibres with Potato Starch *Fibres & Textiles in Eastern Europe* **18** 82

Energy saving due to natural ventilation in housing blocks in Madrid

RA González-Lezcano[1] and S Hormigos-Jiménez[1]

[1]Department of Architecture and Design, Escuela Politécnica Superior, Universidad CEU San Pablo, Urbanización Montepríncipe, 28668 Boadilla del Monte, Madrid, Spain.

Corresponding author's e-mail: rgonzalezcano@ceu.es

Abstract. Getting a healthy and comfortable indoor environment in homes in southern Europe is a complicated task. In continental climates, with very cold temperatures in winter and very hot in summer, energy consumption greatly increases with air conditioning significant spending. To propose action guidelines for use of natural ventilation and to develop effective design strategies is essential. Therefore, and given a specific building type block of flats in Madrid, this article focuses on establishing what periods of the year natural ventilation is required to reduce energy consumption in air conditioning, also considering the quality of the outdoor environment and the design of the building. To develop this, a statistical study of the chosen type, that allows studying the direction and the wind speed in the area, is performed. Analysis of wind pressures in holes in the facade is performed by means of numerical simulations of fluid flow (CFD) inside to later infer in the natural ventilation rate required within policy parameters. With the data obtained, a study of energy saving is made as a function of natural ventilation rate established for the building type.

1. Introduction

1.1. Natural ventilation and its influence on energy saving

Natural ventilation is the process through which indoor air is replaced by fresh outdoor air. This happens through openings in façade in order to achieve users' thermal comfort and good levels of indoor air quality without consuming energy. Natural ventilation is widely used to improve Indoor Air Quality (IAQ) as it affects occupants' health. Good Indoor Air Quality is defined as air free of pollutants although, if it is not good enough, it can cause users irritation in eyes, nose and throat; discomfort; or even health problems such as respiratory or cardiovascular diseases [1]. In this context, air exchange rate is as an extremely important factor to take into account to avoid pollutants concentrations indoors. Air exchange is achieved through ventilation in order to ensure a healthy indoor environment [2-5]. However, the use of insulation and tighter buildings are more which have reduced heating and cooling energy demands. This is to the detrimental of ventilation as air exchange decreases and pollutant concentration indoors increases [6]. Ventilation is included within Heating, Ventilation and Air-Conditioning systems (HVAC), which suppose most of the energy consumption in a building; on the one hand, Orme [7] established in 1991 a percentage of 26.48% of the total annual energy delivered in service and residential buildings; on the other hand, The International Energy Agency ensures that the European HVAC consumption has increased meaningfully, it supposes approximately 50% of the total energy consumption in European buildings [8-9]. In addition, inappropriate ventilation supposes up to 70% in heat losses [5, 10].

Renewable energy is intended to satisfy the energy demand in housing blocks but, in most cases, this is not satisfactory enough. Energy consumption must been avoided wherever it is possible. Therefore, the use of natural ventilation becomes essential to achieve this goal.

1.2. European research on ventilation and regulations regarding Indoor Air Quality in Spain

Although there are regulations available on air quality, there is still a lack of experience and know-how regarding this field, especially within the main stakeholders such as engineers, architects, housing corporations, product producers [11]. This lack of knowledge was identified by several European projects such as the NatVent Project [12] or the HealthyAir Project [13]. Considering the outcomes of the European Project HealthyAir, improvements regarding Indoor Air Quality and ventilation must be developed in three fields [13]: education and awareness; regulation and policies; and research and development. This Project has demonstrated that people in charge of building design are generally not aware of ventilation issues and they declare that they do not have simple tools for creating a proper design regarding ventilation. It is therefore necessary further research and transfer of knowledge.

HealthyAir Project also shows that the regulations need to be updated [13]. Most of building codes consider CO_2 the predominant pollutant in indoor spaces, but new construction methods and technologies have increased the concentration of other pollutants or have crerated new ones.

There are several regulations which can be applied in residential buildings in Spain: the European standard UNE EN 12207:2000 [14]; and the national standards CTE-HS3 [15] and RITE [16]. The national standard CTE-HS3 is a section of the Spanish Technical Building Code which covers the aspects related to Indoor Air Quality in buildings and it has applicability in residential buildings and their different areas. This standard shows the minimum ventilation flows per area (table 1).

Table 1. Minimum ventilation flows required. CTE-HS3

Residential building area	Minimum ventilation flow (L/s)		
	Per occupant	Per usable m^2	Considering other parameters
Bedrooms	5		
Living rooms	3		
Bathrooms and toilets			15 per room
Kitchens		2	
Storerooms		0.7	
Garages			120 per parking space

The other national standard, RITE (Regulation for Thermal Installations in Buildings) establishes which must be the thermal conditions regarding indoor design in buildings in Spain (table 2).

Table 2. Indoor thermal conditions. RITE.

Season	Temperature (ºC)	Relative humidity (%)
Summer	23-25	40-60
Winter	21-23	40-50

1.3. Environmental conditions affecting ventilation in Madrid

Natural ventilation must be an important factor to be taken into account for the building designers. Location, position and design of the building, as well as the environmental conditions, must be studied to reach energy saving through optimum natural ventilation. Natural ventilation of a building occurs because of the ambient condition around it; wind direction and temperature vary along the day thus, a good design focused on natural ventilation must study those changes [17]. The design which takes climate into account is the bioclimatic design and it has the purpose of reducing energy demands for heating and cooling [18]. In this type of design, natural ventilation is included next to passive climate control strategies and renewable energy in bioclimatic design.

The city of Madrid is included within the Community of Madrid, in Spain, in Southern Europe. There are several types of climate in the Community of Madrid [19] as the result of various Atlantic storms with stable air of subtropical origin. Stable air predominates during 60 per cent of the days of the year. Most of the territory is under the Mediterranean climate (temperate climate), which is characterized by dry and hot summers, with an average of temperature above 22ºC, and wet and cold winters. In the urban area of Madrid, the climate is modified by the heat island effect. Heat generated from the constant urban activity causes an increase in average temperature values in both day and night temperatures. Additionally, absolute humidity is lower because of the scarce vegetation and little evaporation. In the city of Madrid, the following climate values, provided by the State Meteorology Agency [20] (AEMET), can be found (table 3), from May to September.

Table 3. Normal climate values in Madrid.

Period: 1981-2010										
Area: Retiro; Latitude: 40º 24' 43'' N; Longitude: 3º 40' 41'' O; Height (m): 667										
Month	T	TM	Tm	R	H	DR	DN	DF	DH	I
May	16.7	22.2	11.3	50	53	7.3	0.0	0.1	0.0	268
June	22.2	28.2	16.1	21	44	3.4	0.0	0.0	0.0	315
July	25.6	32.1	19..0	12	38	1.7	0.0	0.0	0.0	355
August	25.1	31.3	18.8	10	41	1.7	0.0	0.0	0.0	332
September	20.9	26.4	15.4	22	50	3.3	0.0	0.2	0.0	259

T, average monthly/annual temperature (ºC)
TM, average monthly/annual temperature during the day (ºC)
Tm, average monthly/annual temperature during the night (ºC)
R, average monthly/annual precipitation (mm)
H, average relative humidity (%)
DR, average monthly/annual number of rainy days equal or above 1mm
DN, average monthly/annual number of snowy days
DF, average monthly/annual number of foggy days
DH, average monthly/annual number of days of frost
I, average monthly/annual number of sunny hours

As it can be seen in the table 3, natural ventilation is possible in most of the months along the year, whenever outdoor temperature does not exceed 28ºC [21]. In addition, the Outdoor Air Quality [22-23] in Madrid has appropriate values to develop ventilation in buildings, considering values provided by AEMET related to the concentration (µg/m3) of NO2, NO, CO, O3 in the environment (table 4).

Table 4. Outdoor Air Quality in Madrid

Element	NO_2	NO	CO	O_3	PM_{10}
Concentration (µg/m^3)	50-75	10-20	100-200	100-120	20-31

It is difficult to quantify wind velocity and direction, as it is varying continuously and it also depends on the area where anemometers are located. It has been chosen a specific area in Madrid, which is called Retiro (Latitude: 40º 24' 43'' N; Longitude: 3º 40' 41'' O; Height (m): 667), to obtain the average values [24-25] of the wind that are going to be considered for this study (figure 1). Wind speeds above 5m·s^{-1} have a frequency of occurrence much lower than the ones below that value; therefore, the wind speeds considered in this study are: 0m·s^{-1} for wind speeds below 1m·s^{-1}; 2m·s^{-1} for wind speeds of 1-3 m·s^{-1}; and 4 m·s^{-1} for the rest of the values below 5 m·s^{-1}.

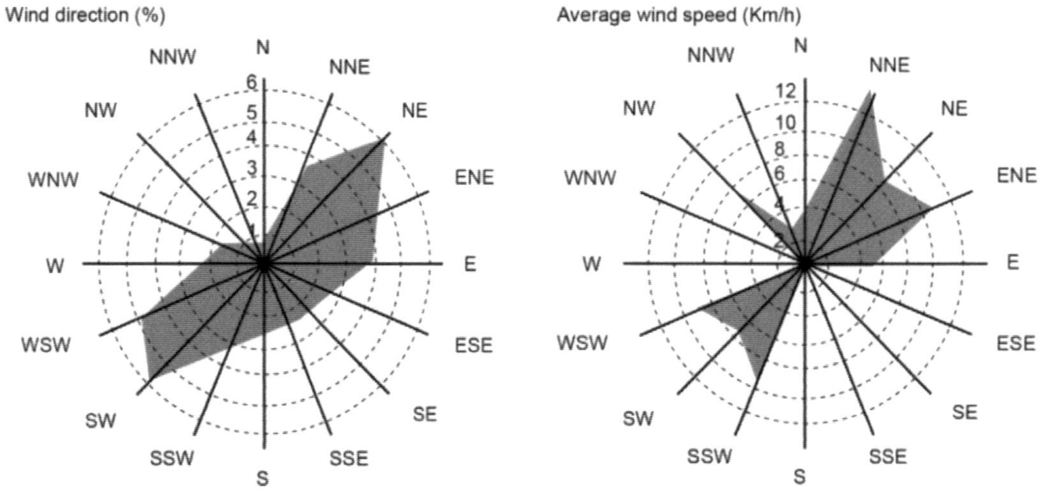

Figure 1. Wind direction (%) and average wind speed (Km/h) in Madrid

1.4. Spending per household on electricity

Considering a previous study, developed by AIS Group from 2010 to 2013, which measures the evolution of the average expenditure on energy in Spain; every home in the country spends an average of €725 for paying the electricity bill, although in most areas of the city of Madrid the electricity consumption is lower than €650 (figure 2).

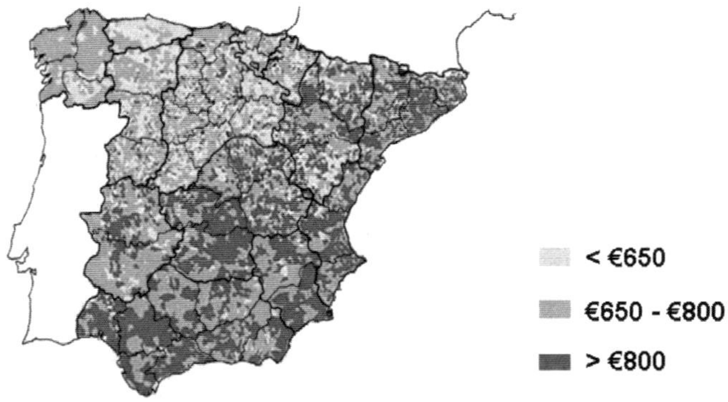

Figure 2. Average expenditure per household on electricity in 2013.

Air-conditioning in Madrid represents an average annual consume (per home) of 1220 KW/h per year and an annual cost of €246 per year. This data means that, considering the use of air-conditioning during the months from May to September, families in Madrid spend an average of €49.2 per month.

1.5. Social Housing in Madrid

The price per square meter of Private Market Housing was € 2,085.5in 2007; it was € 1,014.6higher than Social Housing. From 2013 until nowadays, prices of Private Market Housing (figure 3) in Spain have been reduced in such a way that differences between both types are about 300€ [26].

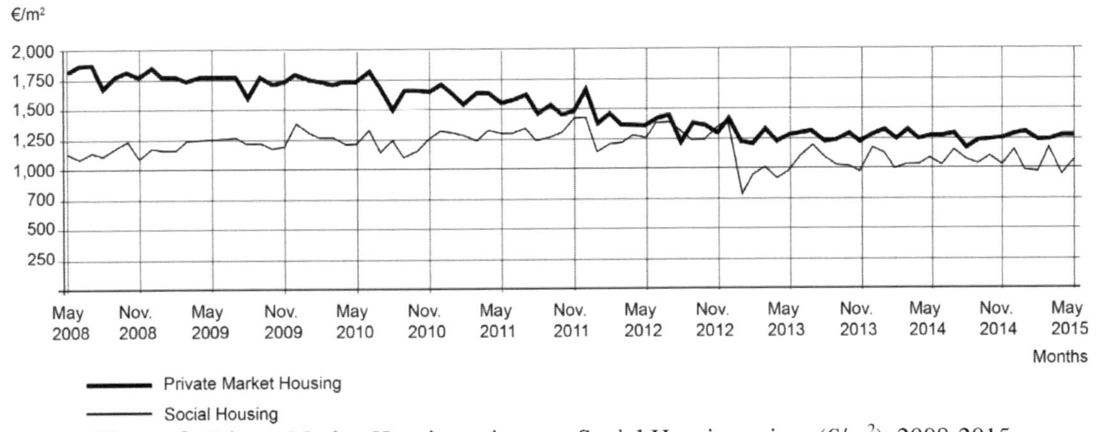

Figure 3. Private Market Housing prices vs. Social Housing prices (€/m^2). 2008-2015

People living in Social Housing Blocks usually have low income. Thus, providing different ways of economic saving reducing energy consumption supposes a benefit for them. What is intended to achieve in this paper is to acquire comfort conditions and good Indoor Air Quality in the areas of the home where occupants spend most of their time, bedrooms and living rooms, using natural ventilation whenever it is possible addressing therefore the minimum energy consumption.

1.6. Objectives for this study

Considering the information provided, the main objectives for this study can be resumed as the following:

- Determine ventilation rates and air changes per hour in living rooms and bedrooms
- Define the stages when cooling by air conditioning can be replaced by natural ventilation
- Determine both energy saving and economic saving that supposes the use of natural ventilation

2. Methodology

2.1. Previous data

- Wind speeds considered for this study:
  - $0 m\cdot s^{-1}$ for wind speeds below $1 m\cdot s^{-1}$
  - $2 m\cdot s^{-1}$ for wind speeds of 1-3 $m\cdot s^{-1}$
  - 4 $m\cdot s^{-1}$ for the rest of the values below 5 $m\cdot s^{-1}$
- Wind directions: Northeast and Southwest
- Indoor comfort conditions:
  - Summer: temperatures of 23-25 °C and relative humidity of 40-60%
- Minimum ventilation flow
  - Bedrooms: 5l/s per occupant
  - Living rooms and dining rooms: 3l/s per occupant
  - Considering the study of C. Dimutroulopoulou and J. Bartzis [27] where they determine that several health problems can started to be developed by the occupants of indoor spaces (irritations or allergies can be suffered) above air flows of $10 L\cdot s^{-1}\cdot person^{-1}$, a maximum threshold of $10 L\cdot s^{-1}\cdot person^{-1}$ has been established for natural ventilation air flow indoors.
- Periods of the year when people uses air-conditioning [21].

- Bedrooms: 13:00-7:00 (18 hours per day)
- Living rooms: 13:00-22:00 (9 hours per day)

2.2. Morphologic characteristics of the housing block studied
In Spain, the most common typology of housing is the 6-7 storey building. Linear housing blocks with rectangular shape in floor plan. This paper have considered the next type of housing block to develop the study located in Madrid [29], oriented to Southeast (figure 4).

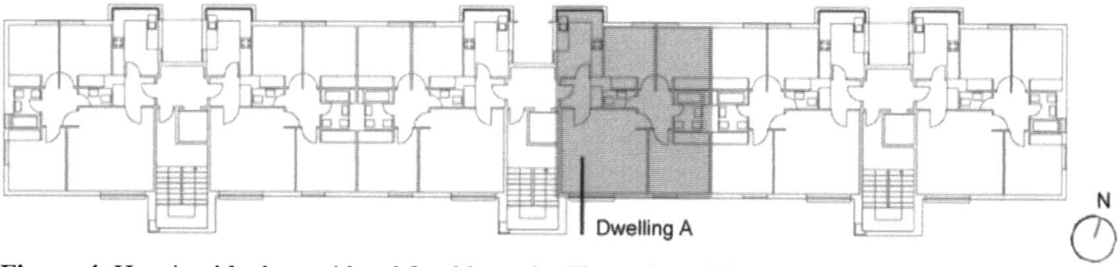

Figure 4. Housing block considered for this study. Floor plan of 6 storey building located in Madrid

Dwelling A highlighted in figure 4 represents the floor type of every dwelling in the housing block and it is detailed in figure 5. It has a total of 67.53 m^2. The living room and bedroom 3 face the southeast façade and the rest of bedrooms (bedroom 1 and bedroom 2) are located in the northwest area of the dwelling. The rest of dwellings are equal to dwelling A o symmetric to it.

Figure 5. Floor plan of dwellings in the housing block. Dwelling useful floor area: 67.53 m^2

2.3. Air flow simulation
To develop methodologically this work, previous studies have been taken into account. Bangalle et al. [17] developed flow simulation considering wind pressure differences as the main tool for natural ventilation; this occurs in buildings through openings in façade causing a circulation of flow due to differences in indoor and outdoor temperatures and pressure differences in façades (faces) of the building. This is why windows (and the influence on one into other) play an important role in natural ventilation processes. Even if there is no external wind, natural ventilation in rooms can occur because of recirculating flow due to temperature differences. However, as it has been shown, natural

ventilation can be much more effective under certain climate conditions. Having an appropriate outdoor environment, with an adequate outdoor air quality, is essential [28].

Different approaches can be established to study ventilation flows. The most important ones are experimental studies, empirical models and numerical simulations by computational fluid dynamics (CFD). CFD simulations have several advantages such as the reduction of costs. Different models can be also considered modifying the architectonic design to predict which the best one in terms of ventilation is. In this study, air flow simulation has been carried out using the CFD package called FLUENT in order to predict wind pressures in façade.

3. Results

3.1. Natural ventilation rates in rooms

CFD simulations have been developed [21] within a flow rate domain with the next dimensions:

- Length: 22H; width: 16H; height: 3H

Being H the height of the building which has been considered for this study (19m). The right-side and left-side limits all along the length of the domain have been established as the inlet and outlet planes of the wind flow through the flow domain. The symmetry planes considered are the top part of the flow domain and the rest of the sides; therefore, normal speeds values are zero in those planes.

CFD simulations have been based on k-epsilon turbulence model and the Quick scheme and Simplec algorithm have been used in order to obtain more quickly a converged solution.

CFD simulations have been developed to obtain wind pressures in openings in the. The studied block is an 8-storey building, which has been considered in its whole to obtain the pressures. The cross ventilation rates can be obtained by modeling a single storey of the building, taken into account its location height, and by using the software COMIS. In every floor there are 24 openings (windows) in the façade, where wind pressures have been assessed using CFD predictions. Figure 6 shows wind pressure coefficients in floors between the first one and the fourth one (4F) and from the fifth one and the eighth one (8F) considering the wind directions from the South direction.

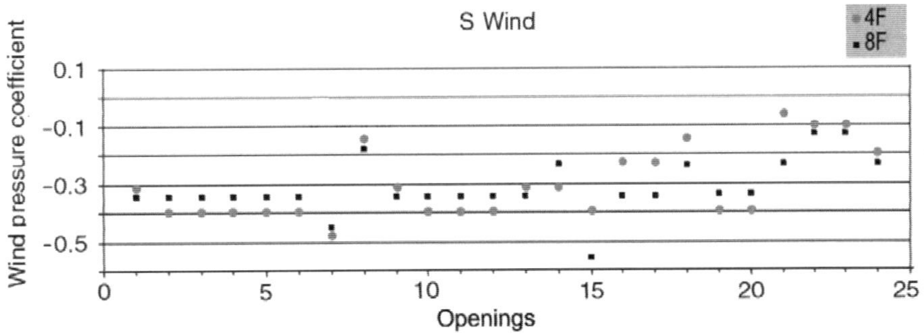

Figure 6. Wind pressure coefficient in the South façade with Southwest wind

It is important to take into account the vertical variation of the wind speed all along the height of the building, having as reference point the terrain. In COMIS, this variation is established by the use of power law model. To simulate the air flow, the process was simplified to reduce time for the converged solution but acquiring correct outcomes, this is why, in view of that network model which was developed had into account all the rooms of a single floor, air flows between the different storey of the building were not considered nor infiltrations by the façade through cracks. The model in COMIS includes 68 flow links which represent the windows and the doors in the whole floor and also 24 nodes in the external façade, one per window of each floor for which a wind pressure coefficient (obtained by the CFD simulation) is assigned.

3.2. Energy saving by means of natural ventilation

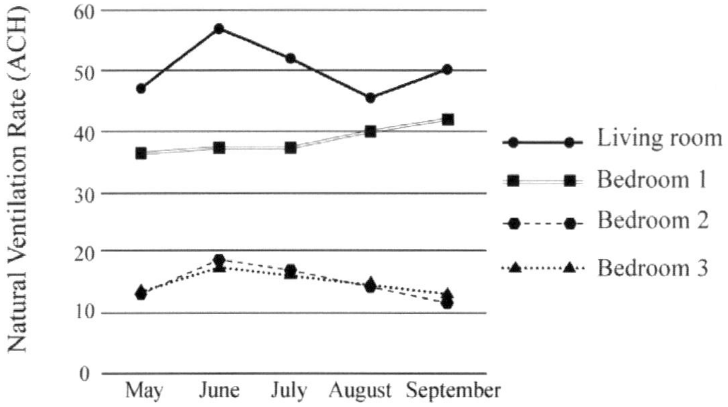

Figure 7. Mean natural ventilation rates reached in living rooms and in bedroom in the housing block

The average natural ventilation rates have been calculated through the method shown above per each of the rooms where air conditioning is installed (living rooms and bedrooms) in both 4F and 8F group of floors. Figure 7 shows the average natural ventilation rates determined for living rooms and bedrooms of the flats 8F (floors between 5 and 8) of the studied housing block. It can be seen that, in some cases, air changes above 50ACH can be reached in some rooms and in every case the current regulation about ventilation is met. This study has shown that indoor temperatures suffer a wide variation when air changes per hour go from 2ACH to 30ACH. Figure 9 illustrates the energy saving that supposes natural ventilation use (avoiding air-conditioning use).

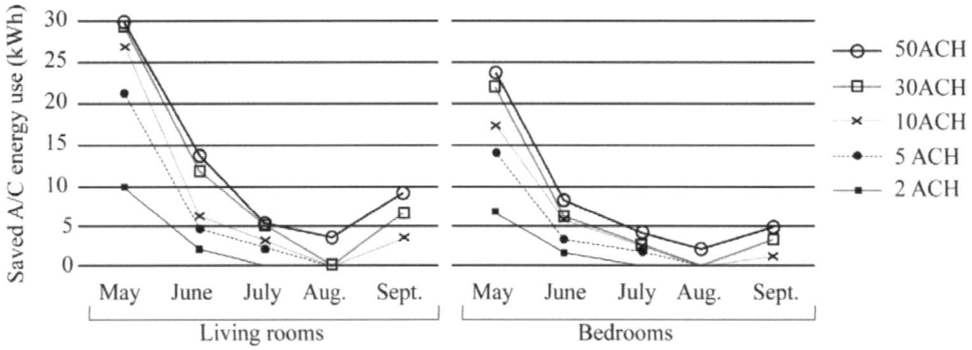

Figure 8. Saved air-conditioning energy by means of natural ventilation in living rooms and bedrooms

Figure 8 shows that the saved air-conditioning energy increases with the natural ventilation rate. Only with a ventilation rate of 2ACH a high saving can be reached in both months May and September but this saving is very low during June, July and August. By increasing the natural ventilation rate, the energy saving also increases, until values of 30ACH are reached, from which the growth is no longer significant. The outcomes shown in figures 7 and 8 can be explained as follows:

- In the Mediterranean climate, in wild months such as May (shown in table 3), outdoor temperature is below 26ºC. Low ventilation rates during this month will maintain indoor temperature under 26ºC therefore, air-conditioning energy can be avoided.
- When temperatures during the day start to be closer to 26ºC, it is necessary to increase air change rates to obtain a higher natural ventilation.
- When outdoor temperature has values of 26ºC or higher, the use of natural ventilation shall be limited, it will be used whenever outdoor temperature is below 26ºC. As is can be seen in figure 8, the saving values in the hottest months (July and August) are minimum compared to the rest of the months considered.

Figure 9 in percentage of saving during the months concerned. This analysis shows that, by using natural ventilation, a percentage of 13% of annual saving in air-conditioning energy use is achieved.

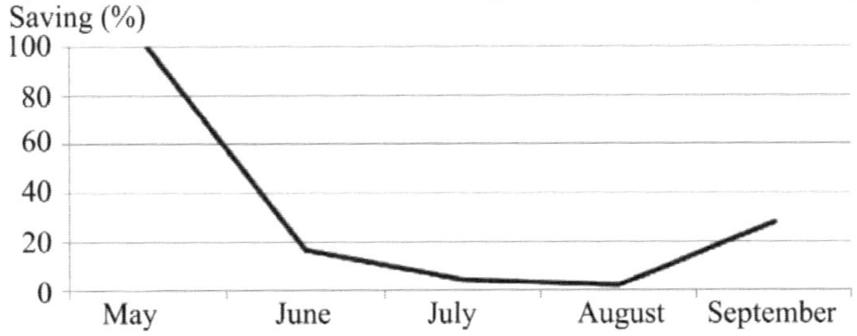

Figure 9. Economic saving percentage (%) in every month by means of using natural ventilation

4. Conclusions

In this study a practical approach and its application is presented and exemplified with a case-study in which natural ventilation rates achievable in living rooms and bedrooms are shown, as well as the minimum energy saving that can be reached by means of using natural ventilation in social housing in a widely used architectonic typology in Spain for this kind of building.

The modeling approach involves the use of CFD simulation software for predicting wind pressures on the windows on the façade of the building, using a network model to predict natural ventilation flow rates in residential flats. Cooling energy saving achievable was taken as the use of cooling energy that can be avoided due to the presence of natural ventilation during the periods in which the outdoor environment permits its use, so that the indoor temperature can be maintained below the tolerable upper limit (26ºC) through natural ventilation without using air-conditioning energy. However, within a nominal air conditioning period, once the indoor temperature exceeds the tolerable limit and air conditioning has been turned on, the assumption has been made that the air conditioners were maintained working during the remaining nominal air-conditioning period. The results reflect that natural ventilation can be highly effective to reach optimal thermal comfort conditions by reducing the use of air-conditioning, meaning a 13% of saving considering the total cooling energy use.

Acknowledgments

The authors would like to express their gratitude to Rebirth Inhabit Research Group of Escuela Politécnica Superior and to Universidad CEU San Pablo for the facilities sand resourced provided.

References

[1] Liddament MW, 1996, A Guide to Energy Efficient Ventilation, 2nd ed., Air Infiltration and Ventilation Centre, Berkeley, 254.

[2] Elkilani A, Bouhamre W, 2001, Estimation of Optimum Requirements for Indoor Air Quality and Energy Consumption in Some Residences in Kuwait, *Environment International*; **27**: 443-447.

[3] Chela F, Husaunndee A, Inard C, Riederer P, 2009, A New Methodology for the Design of Low Energy Buildings, *Energy Build*; **41**: 982-990.

[4] Kavgic M, Mumovic D, Stevanovic Z, Young A, 2008, Analysis of Thermal Comfort and Indoor Air Quality in a Mechanically Ventilated Theatre, *Energy Build*; **40**: 1134–1343.

[5] Joo J, Zheng Q, Lee G, Kim JT, Kim S, 2012, Optimum Energy Use to Satisfy Indoor Air Quality Needs. *Energy Build*;**46**:62-67.

[6] Kukadia V, Palmer J, 1998, The Effect of External Atmospheric Pollution on Indoor Air Quality: a Pilot Study, *Energy Build*; **27**: 223-230

[7] Orme M, 1991, Estimates of the Energy Impact of Ventilation and Associated Financial Expenditures. *Energy Build*; **33**: 199-205

[8] Perez-Lombard L, Ortiz J, Pout C, 2008, A Review on Building Energy Consumption Information, *Energy Build*; **40**: 394–398.

[9] Orosa JA, Oliveira AC, 2009, Energy Saving With Passive Climate Control Methods in Spanish Office Buildings. *Energy Build*;**41**(8):823-828.
[10] Knudsen HN, 2005, HOPE. Health Optimisation Protocol for Energy-efficient Buildings Pre-normative and Socio-economic Research to Create Healthy and Energy Efficient Buildings, 1st ed., Intelligent Energy Europe.
[11] Schulze T, Eicker U, 2013, Controlled Natural Ventilation for Energy Efficient Buildings. *Energy Build*; **56**: 221-232.
[12] Kukadia V, 1998, Work Package 2: Performance of Naturally Ventilated Buildings – Final Monitoring Report, in: NatVent (Ed.), Pan-European Project Encouraging the Use of Natural Ventilation in Office-type Buildings.
[13] Bluyssen PM, de Richemont S, Crump D, Maupetit F, 2010, Actions to Reduce the Impact of Construction Products on Indoor Air: Outcomes of the European Project;19(3):327-339.
[14] Ministerio de fomento. Código Técnico de la Edificación. Documento Básico HS. Salubridad.
[15] Ministerior de Industria, Energía y Turismo. RITE.
[16] Bangalee MZI, Lin SY, Miau JJ, 2012, Wind Driven Natural Ventilation Through Multiple Windows of a Building: A Computational Approach. *Energy Build*; **45**:317-325..
[17] Omer AM, 2008, Renewable Building Energy Systems and Passive Human Comfort Solutions, Renewable and Sustainable Energy Reviews; **12**: 1562–1587.
[18] The physical environment of Madrid. Available at www.madrid.org (Accessed 3 September 2015).
[19] State Meteorology Agency. AEMET.
[20] Yik FWH, Lun YF, 2010, Energy Saving by Utilizing Natural Ventilation in Public Housing in Hong Kong. *Indoor Built Environ*; **19**(1):73-87.
[21] State Meteorology Agency. AEMET. Outdoor Air Quality in Spain. Available at http://www.aemet.es/ (Accessed 3 September 2015).
[22] Particulate Matter (PM10): annual mean concentrations in Europe. Available at http://www.eea.europa.eu/themes/air/interactive/pm10 (accessed 17 April 2015).
[23] Madrid. Climate. Available at: http://www.meteoclimatic.net/ (Accessed 4 September 2015)
[24] Madrid. Retiro. Wind Statistic and climate. Available at: http://es.windfinder.com/ (Accessed 4 September 2015).
[25] Consejo General del Notariado. Centro de información estadística del notariado. Available at: http://www.notariado.org/liferay/web/cien/estadisticas-principales/inmuebles/relacion-vivienda-libre-vs-vpo (Accessed 4 September 2015).
[26] Dimitroulopoulou C. Ventilation in European dwellings: A review.
[27] Social Housing Architecture awards. Available at www.madrid.org (Accessed 20 August 2015).
[28] Chen Q, 1997, Computational fluid dynamics for HVAC: successes and failures, ASHRAE Transactions; **103**: 178-187.

IOP Conf. Series: Materials Science and Engineering **138** (2016) 012003 doi:10.1088/1757-899X/138/1/012003

Method validation for chemical composition determination by electron microprobe with wavelength dispersive spectrometer.

R. Herrera-Basurto[1], F. Mercader-Trejo[2], N. Muñoz-Madrigal[1], J.M. Juárez-García[3], A. Rodriguez-López[2], A. Manzano-Ramírez[4].

[1] Universidad Aeronáutica en Querétaro, Qro., Carretera Estatal 200 Querétaro-Tequisquiapan No. 22154, Colón, Querétaro. México, C.P. 76270.
[2] Universidad Politécnica de Santa Rosa Jáuregui, Carr. Federal 57, km 31-150 QRO-SLP, Santa Rosa Jáuregui, Querétaro, Qro., México 76220.
[3] Centro Nacional de Metrología, km 4,5 carretera a Los Cués, Municipio del Marqués, 76246 Querétaro., México.
[4] CINVESTAV-Qro, Libramiento Norponiente No. 2000, Fraccionamiento Real de Juriquilla, Querétaro, Qro., México 76230.

E-mail corresponding: raul.herrera@unaq.edu.mx

Abstract. The main goal of method validation is to demonstrate that the method is suitable for its intended purpose. One of the advantages of analytical method validation is translated into a level of confidence about the measurement results reported to satisfy a specific objective. Elemental composition determination by wavelength dispersive spectrometer (WDS) microanalysis has been used over extremely wide areas, mainly in the field of materials science, impurity determinations in geological, biological and food samples. However, little information is reported about the validation of the applied methods. Herein, results of the in-house method validation for elemental composition determination by WDS are shown. SRM 482, a binary alloy Cu-Au of different compositions, was used during the validation protocol following the recommendations for method validation proposed by Eurachem. This paper can be taken as a reference for the evaluation of the validation parameters more frequently requested to get the accreditation under the requirements of the ISO/IEC 17025 standard: selectivity, limit of detection, linear interval, sensitivity, precision, trueness and uncertainty. A model for uncertainty estimation was proposed including systematic and random errors. In addition, parameters evaluated during the validation process were also considered as part of the uncertainty model.

1. Introduction

No matter what it costs, any testing and calibration laboratory need to be recognized for the quality of its results. The ISO/IEC 17025 [1] standard establishes the general requirements for the competence of testing and calibration laboratories. Laboratories accredited with ISO/IEC 17025 must comply its requirements grouped in two main fields: (1) Management Requirements; part 4 of the standard, and

(2) Technical Requirements; described in section 5 of the standard. If the laboratory is certified with ISO 9001 [2] standard, it is complying the section 4 of the ISO/IEC 17025. However, section 5 demands the knowledge and experience in specialized topics, such as, fundamentals of metrology, method validation and estimation of uncertainty. Therefore, method validation is a fundamental topic at any laboratory due to the benefits obtained to demonstrate its technical competence and the validity of its results.

On the other hand, wavelength-dispersive X-ray spectrometers (WDS) have been around for more than 60 years. In this time, they have maintained a central role in microanalysis because of their effectiveness in measuring X-rays reproducibly, and because of their high spectral resolution. By the above the WDS has been in the market for many years and it is still the preferred X-ray analytical system for many, mainly due to its qualifications of resolution spectral. For comparison, an energy dispersive spectrometer (EDS) has an energy resolution of around 130 eV (when measured as the full width at half maximum, FWHM, of Mn Kα) as-compared to a traditional wavelength-dispersive spectrometer (WDS), which has an energy resolution of around 16 eV, almost an order of magnitude improvement [3]. The WDS microanalysis advantages can be reinforced having information about validation of the applied method to obtain reliable results. Method validation information is crucial to have a good uncertainty estimation of the result of the measurement [4].

Because there is a lack of information about method validation with WDS microanalysis, authors hope to contribute with novel information. A method for the determination of chemical elemental composition of a binary Au-Cu alloy was studied. Evaluation of the following parameters is reported: selectivity, limit of detection, linear interval, sensitivity, precision, trueness and uncertainty.

2. Materials and Measuring Method

2.1. *Materials*

SRM 482 NIST [5] is a set of 6 wires with different chemical elemental composition of the binary Au-Cu alloy. There are two pure metallic wires, one of Au at 100 % and another one of 100 % Cu and four binary alloys Au-Cu with chemical nominal compositions as follows expressed as mass fraction (%):Au20-Cu80, Au40-Cu60, Au60-Cu40 and Au80-Cu20. The SRM 481 standard was arranged at the same capsule as a condition of practice good laboratory. The alloy Au80-Cu20 was considered as test sample, while the another alloys were used as standard and control sample.

2.2. *Equipment*

All experiments were done using an electron probe micro analyzer (EPMA) model Super-Probe JXA 8200 JEOL equipped with 3 WDS and one EDS. WDS are from JEOL, each one containing 2 analyzer crystals. JXA-8200, has one EDS Si (Li), window Be Thermo Noran, 130.8 eV at 5.9 keV. Specimen tilt 0°, take off angle 40°, EDS elevation angle 40°, EDS azimuth angle 140°, insertion distance 87 mm, height 11 mm.

The microprobe is controlled with the software JXA-8200 JEOL V01.02. Record and analysis of the data was done with the same software. The EPMA has a UNIX operating system; therefore, it is not possible to obtain spectral files in emsa format.

3. Results and Discussion

3.1. *Measurand definition*

The basis for quantitative X-ray microanalysis is that, to a first approximation:

$$\frac{C_{sample}}{C_{std}} \infty \frac{I_{sample}}{I_{std}} \cong k\ value \qquad (1)$$

Where C is the mass (weight) concentration of an element. *I* is the measured characteristic X-ray intensity for the element (corrected for background, peak overlap and dead time). The subscript "sample" denotes the concentration and intensity of the element in the sample, and "std" refers to the standard of known concentration, typically it is a pure element. The ratio of the characteristic intensities measured on the standard is known as the "*k* ratio" or "*k* value". The proportionality sign in equation (1) indicates that the relationship between concentration and characteristic X-ray intensity is not exact. There exist "matrix effects", arising from the nature of the electron and X-ray interactions with matter, which modify the measured intensities, and which depend on the unknown composition of the sample. A variety of approaches are used (ZAF, phi(rho)-z, empirical) to calculate correction factors for these matrix effects.

3.2. Selectivity

The selectivity of a method is usually investigated by studying its ability to measure the analyte of interest in samples to which specific interferences have been deliberately introduced (those thought likely to be present in samples) [6].

The selectivity of the method was evaluated through the ability of the spectrometer to change the analyzer crystal for the analysis of the same element due to the use of different acceleration voltages: 5 and 25 kV, table 1. The interferences were avoided through the selection of X–ray lines to different level for each element. Table 1 shows the experimental conditions used for the analysis of each one of the elements including type of analyzer crystal, acceleration voltage and working distance. Measurement results of the Au80-Cu20 alloy, taken as a sample, are also included in table. The Au80-Cu20 standard was measured at 5 and 25 kV electron beam. Under these conditions, 7 microanalyses with duration of 40 s each were recorded. The selectivity of the WDS microanalysis method for Au-Cu alloy was through of the estimation of the variation coefficient or relative standard deviation (RSD) 0.49 to 0.7 % to Au and 0.67 to 0.76% to Cu. Thus, we demonstrated the ability of the method to confirm analyte identity and its ability to measure the isolated analyte and with interferences for the effect of the binary alloy, taken as a sample, are also included in table 1

Table 1. Experimental conditions and results for the evaluation of selectivity.

Acceleration voltage	Au80-Cu20 @ 5 kV, 50 μA		Au80-Cu20 @ 25 kV, 17 nA	
Analyzer crystal	PET	TAP	LiF	LiF
Element-X-ray line	Au-Mα	Cu-Lα	Au-Lα	Cu-Kα
Counts	452592.5	279216.8	215934.7	176421.2
% RSD	0.49	0.67	0.7	0.76

3.3. Limits of detection and quantification

Limit of detection (LOD) was based on the analysis of blank samples following the whole measurement procedure. Calculation of the LOD was done through the measurement of pure metallic wires. Pure Cu wire was used for determining the detection limit of Au, considering this wire as a sample blank for Au. Whereas for determining the detection limit of Cu, pure Au wire was measured. Fourteen independent measurements of the sample blank were made under repeatability conditions.

As suggested by the Eurachem method validation working group [7], the LOD was calculated using equation (2) through the standard deviation of the measurement of 100 % Cu standards and 100 % Au.

$$LOD = 3\, s_0' \qquad (2)$$

where:

$$s_0' = \frac{s_0}{\sqrt{n}} \qquad (3)$$

s_0 - is the estimated standard deviation of n single results at or near zero concentration.
s'_0 - is the standard deviation used for calculating LOD.
n – is the number of replicate observations.

Table 2 shows information considered in the calculation of the method detection limit. Signal to Noise (S/N) ratio is also included.

Table 2. LOD and LOQ information for Au and Cu.

25 kV, 17 nA									
Au-Lα					Cu-Kα				
Signal (S)	Noise (N)	Ratio S/N	LOD	LOQ	Signal (S)	Noise (N)	Ratio S/N	LOD	LOQ
291301.7	2963.3	98.3	0.87	0.23	740750.1	727	1018.9	1.15	0.33

Limit of quantification (LOQ) is a parameter associated to the LOD. LOQ is the lowest level of analyte that can be determined with an acceptable level of confidence. In practice, according to the Eurachem working group, LOQ is calculated by most conventions to be the analyte concentration corresponding to the obtained standard deviation (s'_0) at low levels multiplied by a factor, k_Q. The IUPAC default value for k_Q is 10 [8] and if the standard deviation is approximately constant at low concentrations this multiplier corresponds to a relative standard deviation (RSD) of 10 %.

3.4. Working Range

In order to establish the instrument working range, Au-100, Au60-Cu40, Au40-Cu60, Au20-Cu80 and Cu100 standards were measured 10 times on the surface of each wire. Measurements were obtained with an electron beam of 5 and 25 kV respectively. The average value of the measurements was used to build the response curve. Figure 1 show the calibration function obtained through four calibration points. A high correlation factor (r) above than 0.99 was obtained indicating the fitness for purpose of the calibration curves. Parameters of the least square linear calibration curve are summarized in figure 1 and table 3.

The working range was assessed with a calibration function. The calibration curve for Au has a slope lower than Cu for both conditions (5 and 25 kV). Thus, Cu measurements point out more sensitivity that the Au measurements.

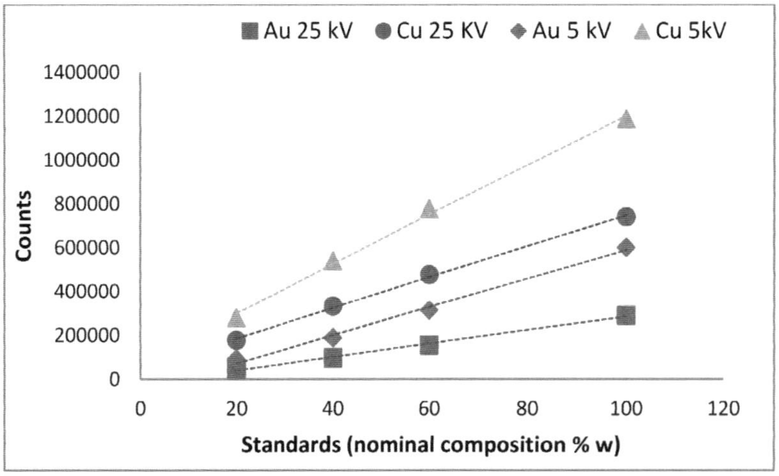

Figure 1. Performance of the calibration function.

Table 3. Parameters working range

	5 kV, 17 nA		25 kV, 50 μA	
	Au	Cu	Au	Cu
Intercept	-56898	75463	-21167	45994
Slope	6458.1	11276	3067.1	7016.8
Correlation coefficient, r	0.9937	0.9967	0.994	0.9983

3.5. Sensitivity

Analytical sensitivity is the change in instrumental response which corresponds to a change in the measured quantity (for example an analyte concentration), i.e. the gradient of the response curve [9,10]. The prefix 'analytical' is recommended to avoid confusion with 'diagnostic sensitivity' used in laboratory medicine [11].

Sensitivity was estimated as the slope of the response curve in the table 4, shown the results for each element in term of the counts and K value.

Table 4. Sensitivity values for counts and k value

Slope	Au (counts)	K value Au	Cu (counts)	K value Cu
5 kV	6422.31	0.0104	11469.5	0.0095
25 kV	3025.76	0.0106	7041.46	0.0097

3.6. Precision

Approaches to simultaneous determination of repeatability and intermediate precision are described in ISO 5725-3 [12]. In addition, a design based on the harmonized guidelines for single-laboratory validation of methods of analysis [13] offers the possibility to determine repeatability and intermediate precision from a single study. Subsamples of the selected test material are analyzed in replicate under repeatability conditions across a number of different runs, with maximum variation in conditions between the runs (different days, different analysts, different equipment, etc.). Via one-way ANOVA [14,15], repeatability can be calculated as the within-group precision, while the intermediate precision is obtained as the square root of the sum of squares of the within group and between-group precision. This type of design can provide an efficient way of obtaining sufficient degrees of freedom for estimates of repeatability and between-group precision. The precision was quantified through of assessment of the repeatability and reproducibility.

3.6.1. Repeatability Test

The repeatability was based on 8 measurements obtained under repeatability conditions. Repeatability conditions were alloy Au80-Cu20 as test sample, electron beam of 25 KV, beam current 17 nA, distance working 11 mm, the same spot size and assumptions of the high homogeneity of the sample, thus, the analysis was done in different points of the surface of the alloy. Table 5 shows the measurement results under repeatability conditions. The variation coefficient was the parameter use for assessment the measurement repeatability.

Table 5. Repeatability results as variation coefficient

	25 KV	5 KV	25kV	5 kv
	Cu	Cu	Au	Au
	20.65	19.53	79.35	80.47
	20.55	19.13	79.46	80.87
	20.71	18.87	79.29	81.13

	20.55	18.72	79.47	81.28
	20.52	18.61	79.48	81.39
	20.57	18.50	79.43	81.50
	20.60	18.90	79.40	81.10
Averange	20.6	18.9	79.4	81.1
Std Desv.	0.06	0.32	0.06	0.32
Var. Coef.	0.29	1.70	0.078	0.39

3.6.2. Reproducibility Test

The reproducibility assessment was done with an electron beam of 5kV and 25 kV as a reproducibility condition, thus, a change in the beam current was used, analyzer crystal and X-ray line. The test sample was Au80-Cu20 and variance analysis (F test) was applied for reproducibility evaluation. Measurements were performed applying the two different electron beam voltage and the k value data were used in the F test. F test results are shown in the table 6 for each element.

Table 6. Data and F test results as reproducibility

Cu L and K		Au M and L	
5kV	25 kv	5 kv	25 kv
k values	k values	k values	k values
19.53	20.65	80.47	79.35
19.13	20.55	80.87	79.46
18.87	20.71	81.13	79.29
18.72	20.55	81.28	79.47
18.61	20.52	81.39	79.48
18.50	20.57	81.50	79.43
19.53	20.65	80.47	79.35

F test parameters	Cu L and K		Au M and L	
Electron beam	Cu 25 kV	Cu 5 kV	Au 25 kV	Au 5 kV
Average	20.5906667	18.893	79.41266667	81.107
Variance	0.00501467	0.1457628	0.005374667	0.1457628
Data	6	6	6	6
Degrees of freedom	5	5	5	5
F	0.034		0.037	
P(F<=f) one-way	0.001		0.0012	
Value F critical (one-way)	0.198		0.198	

3.7. Trueness

The bias criterion states that if Δv (absolute difference) > Δ_C (uncertainty associated absolute difference), then the difference is greater than can be explained as an instrument correction with a calibration standard. However, if the situation is opposite, spectrometer is accurate and any correction is not required. Equation (4), indicates how to estimate, and Δ_C equation (5) for Δv calculated.

The evaluation of the trueness was realized as a bias (Δ_v). The bias calculated following the equation (5).

$$\Delta_c = (t_{n-1}0{,}95)\left(\frac{s}{\sqrt{n}}\right) + U \qquad (4)$$

$$\Delta v = |\bar{x} - \mu| \qquad (5)$$

Where: μ is the certified value for each element of the sample alloy; $\bar{x}$ is mean of the measurements, s is standard deviation, n is the number of measurement and u is uncertainty standard.

Equations (4) and (5) were considered to obtain the results included in table 7. According to the results, significant bias was not obtained.

Table 7. Criteria bias results

	Δv		Δc	
	Au	Cu	Au	Cu
25 kV	0.25	0.15	0.74	0.76
5 kV	0.29	0.19	0.96	0.94

Of agree with table 7 it not bias was observed, so, the spectrometer fits accuracy.

3.8 Uncertainty

Uncertainty is an interval associated with a measurement result, which expresses the range of values that can reasonably be attributed to the quantity being measured. An uncertainty estimate should take account of all recognized effects operating on the result. The uncertainties associated with each effect are combined according to well-established procedures. Several approaches to obtaining an uncertainty estimate for the results from chemical measurements are described [16, 17, 18, 19, 20]. The uncertainty model approved is the figure 2. The model included source of variation systematic as uncertainty of the standard and the correction for matrix effect with use the software and random sources as repeatability and intermedia precision at figure 3.

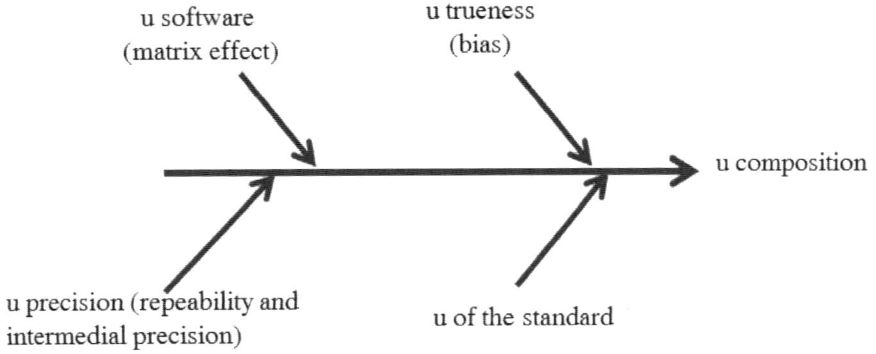

Figure 2. Uncertainty diagram proposed for the determination of element composition of the alloy.

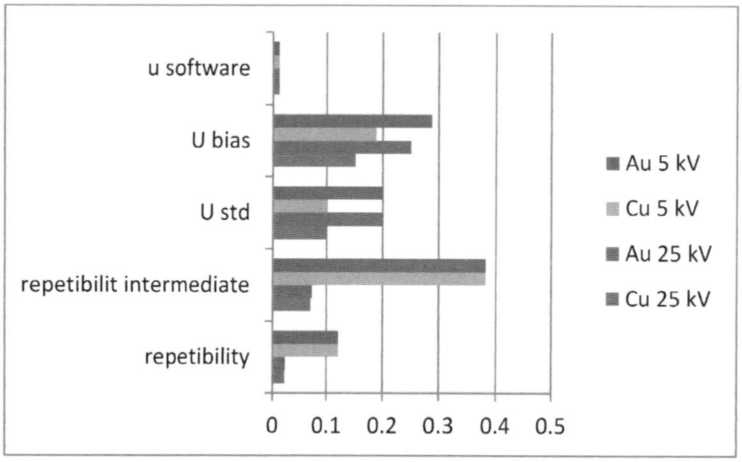

Figure 3. Diagram of the uncertainty budget.

The values for each element composition and their expanded uncertainty are reported in Table 8.

Table 8. Values of the uncertainty expanded for each component

kV	25 KV	5 KV	25kV	5 kv
Element	Cu (kg/kg)	Cu (kg/kg)	Au (kg/kg)	Au (kg/kg)
Value	20.59	18.89	79.41	81.11
Uncertainty expanded	0.39	0.91	0.66	1.07

4. Conclusions

The use of microprobe (EPMA) or electron scanning microscope coupled with WDS has been extended in testing laboratories. Method validation and uncertainty estimation are the main technical requirements to achieve accreditation. This paper shows a way to achieve this goal and it can be considered as a guide for laboratories involved in the accreditation process.

Method validation was performed to determinate the chemical composition by elemental wavelength dispersive spectrometry (WDS). The validation of the method includes uncertainty estimation using a simple model considering and estimation of precision components (repeatability and intermediate precision as source of random variation. The model also includes systematic sources of variation such as the uncertainty due to the software used for matrix effect correction and uncertainty of the standard.

Acknowledgments
R. Herrera Basurto would like to thank to CONACYT, CENAM and UNAQ for the support to develop this work. F. Mercader-Trejo, A. Rodriguez-López and J.M Juárez gratefully acknowledge the SNI and CONACYT for their scholarship.

5. References

[1] ISO/IEC 17025:2005 *General requirements for the competence of testing and calibration laboratories* (ISO Geneva)

[2] ISO 9001:2008 *Quality management systems – Requirements* (ISO Geneva)

[3] Heinrich K F 1981 *beam X-ray microanalysis* ed Van Nostrand (Reinhold Co.)

[4] Mercader-Trejo F, Herrera-Basurto R, de San Miguel E R, de Gyves J 2011 *Int. J. Environ. An.Ch.* **91**11 p 1062

[5] Heinrich K F 1971 *Preparation and evaluation of SRM's 481 and 482 gold-silver and gold-copper alloys for microanalysis* (Vol. 260, No. 28) ed US National Bureau of Standards; for sale by the Supt. of Docs (US Govt. Print. Off.)

[6] Eurachem 1998 *The fitness for purpose of analytical methods: A laboratory guide to method validation and related topics* (England Laboratory of the Government Chemist)

[7] Green J M 1996 *Anal. Chem.* **68** 9 p 305A

[8] IUPAC 1995 *Pure Appl. Chem,* **67** p 1699

[9] BIPM 2012 *International vocabulary of metrology – Basic and general concepts and associated* (JCGM 200:2012, www.bipm.org.)

[10] IUPAC 1995 *Compendium of analytical nomenclature* (IUPAC orange book), Available at www.iupac.org. (Accessed 10 October 2015)

[11] Theodorsson E 2012 Bioanalysis **4** 3 p 305

[12] ISO 5725 *Accuracy (trueness and precision) of measurement methods and results – Parts 1-6* (ISO Geneva)

[13] Thompson M, Ellison S L R, Wood R 2002 *Pure Appl. Chem.***74** 5 p 835

[14] Miller J C, Miller J N 2005 *Statistics and chemometrics for analytical chemistry* ed Pearson (Person Harlow Education)

[15] Ellison S L R, Barwick V J, Duguid Farrant T J 2009 *Practical statistics for the analytical scientist. A bench guide* ed RSC Publishing (Cambridge)

[16] Magnusson B, Näykki T, Hovind H, Krysell M 2012 *Handbook for calculation of measurement uncertainty in environmental laboratories* ed Nordtest Report TR 537. Available at www.nordtest.info. (Acessed 12 Octuber 2015)

[17] ISO 21748:2010 *Guidance for the use of repeatability, reproducibility and trueness estimates in measurement uncertainty estimation* (ISO Geneva)

[18] Eurolab 2007 *Measurement uncertainty revisited: Alternative approaches to uncertainty evaluation* (Technical report No. 1/2007). Available at www.eurolab.org. (Accessed 9 September 2015)

[19] Ritchie N W, Newbury D E and Davis J M 2012 *Micros. and Microanal.* **18** 04 p 892

[20] Marinenko R B and Leigh S 2010 (Materials Science and Engineering) (Vol 7 No 1) IOP Publishing p 012017

Accelerometer based solution for precision livestock farming: geolocation enhancement and animal activity identification

G Terrasson[1], A Llaria[1], A Marra[1] and S Voaden[1]

[1] Ecole Supérieure des Technologies Industrielles Avancées (ESTIA), Technopole Izarbel, 64210 Bidart, France

Email: g.terrasson@estia.fr

Abstract. The rapid evolution of electronics and communication technologies in the last years has contributed to the expansion of Precision Livestock Farming applications. In this context, animal geolocation systems applied to extensive farming are interesting for farmers to optimize their daily work organization. Nevertheless, the deployment of these solutions implies several technical challenges which must be resolved, mainly the energy consumption and the suitability of the communication protocols. A recently developed solution that deals with these technical challenges is the e-Pasto platform, which is composed of low power geolocation devices embedded into collars that offer an energetic autonomy of at least seven months, completed with a visualization user interface. The autonomy is assured employing a duty-cycle operation that results in one geolocation position measurement per hour. This work studies the employ of accelerometers to overcome this limitation assuring, at the same time, the required autonomy for the geolocation device. The authors also propose an algorithm that processes the acceleration data in order to identify the steps of an individual. The whole solution (step identification and geolocation) has been validated by means of several experimental tests.

1. Introduction

From several years, Precision Livestock Farming (PLF) applications have known an increasing deployment mainly due to the evolution of the Information and Communication Technologies and to the miniaturization of the electronic devices. In this frame, the organization of farmers' daily work could be optimized by means of tools capable of locating their animals in real time over large area surfaces. In fact, this kind of applications offers to the farmer the capacity to survey remotely its cattle avoiding unnecessary displacements and, in addition, to define virtual fences in order to better manage the pasture resources.

Nowadays, a significant number of geolocation solutions are based on electronic devices, embedded into collars carried by animals, combined with wireless communication networks that transmit the collected data. Nevertheless, the application of these systems to extensive farming in high mountain pastures implies some technical challenges which must be resolved, mainly the energy consumption and the suitability of the needed communication protocols. In this context, this paper presents an accelerometer based solution that improves the energy efficiency of embedded geolocation devices, allowing at the same time a basic animal activity identification.

The document is structured as follows. Section 2 presents the main technical challenges of animal geolocation in mountain pastures and the e-Pasto platform. The proposed accelerometer based solution is presented in Section 3, along with the obtained experimental results. Finally, Section 4 concludes this paper.

2. Animal geolocation in mountain extensive farming

2.1. Technical challenges
Geolocation embedded electronic devices, usually carried by the animals, and wireless communication networks that transmit the collected position data to the final user, are the basic technologies needed to implement an animal geolocation system. If the geolocation procedure must be performed in large area mountain pastures, applied to extensive farming, there are a number of technical challenges to be resolved.

The main constraint is the energy autonomy of the embedded devices, mainly impacted by the Global Navigation Satellite Systems (GNSS) which are relatively power-hungry. Several methods can be applied to overcome this problem [1, 2], such as increasing the battery capacity (and thus the size and weight of the device), energy harvesting technologies (which are not yet very mature) or the application of a duty-cycle strategy to measure the geolocation positions. This last method obliges to fix a trade-off to assure both the highest autonomy and the maximum number of collected positions. Therefore, the main drawbacks of duty-cycle strategy, considering the autonomy constraint, could be the reduced number of collected position points and the restricted precision.

Concerning the wireless communication networks, the proposed solutions are, in general, based on the existent wireless technologies, such as GSM, WiFi or ZigBee [3]. Nevertheless, in high mountain pastures, it is not always possible to guaranty the correct coverage using these technologies. Moreover, the energy consumption of some of these protocols is excessive for battery powered embedded electronics, limiting even more the energy autonomy of the geolocation devices.

2.2. The e-Pasto platform
An interesting example of animal geolocation system for extensive farming in mountain pastures, which deals, at the same time, with the number of measured positions and the battery autonomy of the devices, is the "e-Pasto" platform [4]. Two main parts compose the e-Pasto solution: the geolocation devices and the visualization user interface.

The geolocation devices, which are embedded into a collar carried by the animals as illustrated by figure 1, determine the position via the GNSS module and, afterwards, the collected positions are sent to a central server by means of the long range low power wireless network developed by SIGFOX©. As shown in figure 2, the visualization interface allows farmers to see the animals' measured positions, as well as other statistical data such as the daily distances covered by each animal. The user can also define virtual fences in the interface to delimit the authorized zones for their animals. If an animal goes out of the authorized area, an alarm message is sent to the concerned farmer. This interface can be accessed from computers or portable devices, such as laptops or smartphones.

Figure 1. Cow carrying the collar with the geolocation device.

Figure 2. Several positions of animals showed in the visualisation user interface.

The correct operation of the overall platform has been validated by means of experimental tests carried out in mountain pastures of France and Spain in 2013 and 2014. The main enhancement provided by the e-Pasto platform concerns the autonomy of the embedded electronic devices, which reaches at least 7 months ensuring, at the same time, the collection of geolocation data in wide surface pastures, along with an innovative long range and low power wireless communication network. Nonetheless, this solution is constrained by its duty cycle operation, which limits the animal position acquisition to one measurement per hour. This duty cycle strategy is very interesting in order to reduce the overall power consumption of the geolocation devices, but it could decrease the interest of users in this kind of systems. Certainly, with this strategy it is not possible to collect geolocation data between two programmed position measurements, even if animals have covered long distances during this period. In addition, the duty cycle implies a continuous position acquisition, procedure which is not optimized to ensure an energy efficient geolocation data measurement especially during the night, when animals stay almost at the same location.

3. Application of accelerometers to improve the geolocation solution

The solution proposed to improve the e-Pasto performances is to incorporate an accelerometer to the geolocation device. In this manner, geolocation positions measurement is only activated when the accelerometer detects a precise movement of the animal, procedure which is more efficient in terms of battery use. The figure 3 compares the principle of duty-cycle operation with the movement detection one proposed in this article.

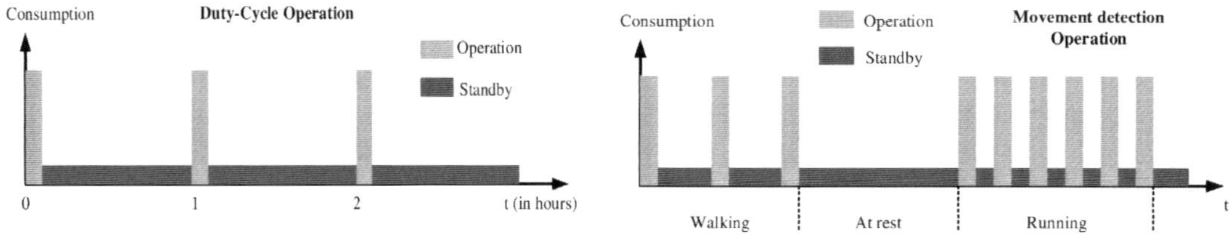

Figure 3. Comparison between duty-cycle and movement detection operation modes.

3.1. Animal activity identification by means of acceleration data analysis

The accelerometer can be employed not only to activate the geolocation measurement, but also to identify the animal activity and its behaviour by analysing the acceleration data [5]. The table 1 compares different algorithms used for individual activity detection by processing acceleration data. According to its accuracy to correctly identify different animal movements combined with its low processing resource requirements and implementation easiness, the most adapted solution for embedded systems seems to be the moving average method. Consequently, this is the algorithm adopted for the proposed solution.

Table 1. Comparison of different acceleration data processing algorithms.

Algorithm	Advantages	Disadvantages
Threshold detection	Low resources consumption Low energy consumption	Low accuracy Impossible to identify the movements
Frequency analysis	Good movement identification Low error margin	Necessary a high frequency sampling High resources and energy consumption
Artificial intelligence	Accurate movement identification Error margin ≈ 0	Long learning phase High resources and energy consumption
Moving average	Acceptable resources consumption Identification of several movements Easy to implement	Not possible to sleep the microcontroller

3.2. Step identification and geolocation with accelerometers: a practical case

In the context of animal geolocation solution improvement, the first stage is to implement the moving average algorithm for step identification. In this way, several experimental tests have been conducted with humans in order to validate the movement identification from the accelerometer data, processed with the moving average algorithm.

The figure 4 illustrates the principle of the implemented algorithm, and the results obtained for one axis acceleration data. At $t=\alpha$, the point *A* of the curve "average t+2" is lower than point *B* of the curve "average t+1". At the same time, the point *B* is lower than point *C* of the curve "average t_0". The distance between *A* and *C* is significant, whereas the distance between *A* et *B* is the same as between *B* and *C*. The measure at α validates the "start" of a step. On the other hand, the measure at $t=\delta$ validates the "end" of a step. At $t=\beta$, the distance between *A* and *C* is not significant enough, and this measure has not been taken into account, which is also the case for $t=\gamma$. This algorithm allows the step detection and, in addition, it can identify, from the distance between the points *A* and *C*, if the individual is walking or running.

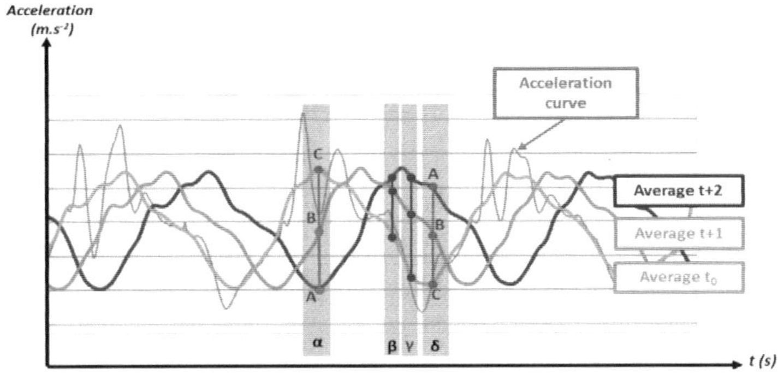

Figure 4. Acceleration curves employed for step detection.

A program that implements the proposed algorithm has been developed and tested with the TD 1204© evaluation electronic card, which is composed of the same elements than the e-Pasto geolocation devices: a low-power microcontroller, a GPS module, and a SIGFOX© modem. In addition to these elements, the TD 1204© also includes an accelerometer, which was not present in the e-Pasto devices. During the experimental tests, the data issued from the accelerometer were processed by means of the proposed algorithm to identify the steps. The GPS module was activated every 50 steps, and the measured geolocation position was transmitted to a data base through the SIGFOX© wireless network. The figure 5 shows in green the geolocation points measured every 50 steps during the tests. The real path, obtained with a classical GPS device of a smartphone, is shown in red to prove the precision of the developed solution. Finally, the figure 6 shows the block diagram of the e-Pasto geolocation device, including the accelerometer.

Figure 5. Geolocation positions obtained every fifty steps.

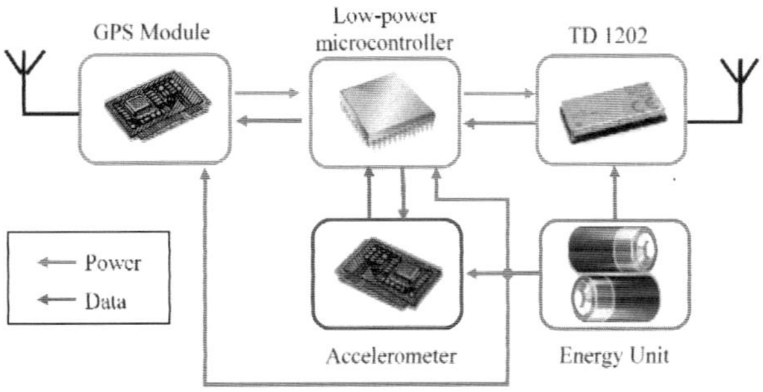

Figure 6. Block diagram of the e-Pasto geolocation device including an accelerometer.

4. Conclusion

This paper has presented a solution, based on accelerometers, that is able to identify the steps of an individual. This technique is useful to collect animal behaviour information and also to use in a more efficiently manner the energy of geolocation devices dedicated to PLF applications. Nevertheless, several aspects must be discussed.

Concerning the impact over the autonomy, the proposed solution improves the energy efficiency because of the optimization of the geolocation positions collection procedure: the GPS module is only activated if a significant movement of the animal is detected by the accelerometer. In addition, the energy consumption due to accelerometers is not very high, and the computational load of the moving average algorithm is moderate, because it is applied to only one acceleration axis. However, the impact of the energy consumption due to both the accelerometer and the processing algorithm over the geolocation device autonomy must be analysed more in detail and, specially, in a high mountain pasture scenario.

The proposed solution has been validated with humans and it is necessary to study its application to animals. The correct positioning of the electronic device over the animal is crucial to achieve the step detection. Related to the interpretation of the acceleration curves, the scientific literature [5] shows that the results obtained for different kinds of animals are similar to those obtained for humans. Finally, it would be desirable to detect, with an adapted algorithm, other kind of animal movements and activities (eating, rumination…), respecting the trade-off with the autonomy. This type of information could be used by the farmers in order to determine the well-being of their cattle.

5. References

[1] Ruiz-Mirazo J, Bishop-Hurley G and Swain D 2011 Automated animal control: can discontinuous monitoring and aversive stimulation modify cattle grazing behavior? *Rangeland Ecology Management* **64** 240–8

[2] Gutierrez A, Dopico N, Gonzalez C, Zazo S, Jiménez-Leube J and Raos I 2013 Cattle-powered node experience in a heterogeneous network for localization of herds *IEEE Trans. Ind. Electronics* **60** 3176–84

[3] Raychaudhuri D and Mandayam N 2012 Frontiers of wireless and mobile communications *Proc. IEEE* **100** 824–40

[4] Llaria A, Terrasson G, Arregui H and Hacala A 2015 Geolocation and monitoring platform for extensive farming in mountain pastures *Proc. IEEE Int. Conf. on Industrial Technology (Seville)* pp 2420–5

[5] Grundy E, Jones M, Laramee R, Wilson R and Shepard E 2009 Visualisation of sensor data from animal movement *Proc. IEEE-VGTC Symposium on Visualization (Berlin)* pp 1–8

Acknowledgements

The work presented in this paper has been developed in the frame of the e-Pasto project (pilot project of AGRIPIR), which has been financially supported by the European Commission in the frame of the Spain-France-Andorra Territorial Cooperation Operational Program (POCTEFA 2007-2013) and the FEDER funds.

IOP Conf. Series: Materials Science and Engineering **138** (2016) 012005 doi:10.1088/1757-899X/138/1/012005

Sensor fusion system to estimate the trajectory of a low cost mobile robotic platform using an Inertial Measurement Unit

J.-S. Botero V., M. Rico G., J.-P. Villegas C.

Grupo AEyCC, Facultad de Ingenierías, Instituto Tecnológico Metropolitano ITM, Carrera 31 No. 54-10, Medellín, Colombia

E-mail: juanbotero@itm.edu.co

Abstract.

In this paper, the development and implementation of an algorithm is presented to identify the trajectory of a mobile robot based on data from an Inertial Measurement Unit with nine degrees of freedom consisting of a 3-axis accelerometers, an 3-axis gyroscopes, an 3-axis magnetometers, and an additional temperature sensor to compensate for temperature errors. The combined information from this set of sensors allows determining the trajectory and orientation of the robot at any moment to complement the information from its navigation system. Initially, it was necessary to build controlled test environments that allow observing and getting to know the trajectory and the dimensions thereof for capturing the raw data of the mobile robotic platform. Then, the data obtained form the sensor was processed off-line applying a Kalman filter with the aim to remove Gaussian noise; to estimating the trajectory and the absolute orientation of the mobile robotic platform the proposed algorithm was implemented. This implementation used the hardware elements are inexpensive, thus allowing the necessary testing, the data analysis and the interpretation of the outcome to be replicable and to be used as an educational tool within courses of undergraduate and master level. The elements of hardware used in this implementation are cheap, allowing its reproduction for the analysis and interpretation of data and can be used as an educational tool in courses of undergraduate and master level.

1. Introduction

The miniaturization of electronic devices, including micromachined sensors [1], and the increase in the processing capacity of low-cost embedded systems have allowed the development of new applications by capturing physical information using processing algorithms in the area of sensor fusion [2, 3]. This fusion allows the derivation of information from measures and reduces the estimation error by combining observations or by integrating algorithms. Particularly in recent years, Inertial Measurement Units have been developed, accompanied by sensor fusion techniques to derive all kinds of mechanical variables used nowadays in almost every mobile consumer device to interpret motion variables. One of the areas of further exploration is the monitoring of human activity [4], [5], [6], sports [7], [8], and to assist in the treatment and diagnosis of many diseases [9], [10], [11]. In addition to global positioning systems [3], [12], or simply as absolute guidance system [13], [14]. All these lines of research make it necessary to progress in sensor fusion algorithms and optimization models to deploy applications in simple systems. Furthermore, the mobile robotic platforms have focused increasingly on daily life applications; it is estimated that in 2013 more than 6 million of those units were in use. Growth in this segment also encouraged the development of an area of research aimed at the design and implementation of navigation systems that enable robots to map the environment and to identify and overcome obstacles to perform their specific task. Until now, most of the operating units derive information about the spatial location and obstacles from encoder signals from the wheels or collisions with objects to complement the inertial measurement system [15], and using Unscented Kalman Filter (UKF) [16], [17] . Although this method has proven effective in domestic applications, it is certain that sensory fusion can optimize currently present problems and eliminate encoders to reduce costs, for this reason we carried out the here described study. In this paper, the development and implementation of an algorithm is presented to identify the trajectory of a mobile robot based on data from an Inertial Measurement Unit with nine degrees of freedom consisting of three accelerometers, three gyroscopes, three magnetometers, and an additional temperature sensor to compensate for temperature errors. The combined information from this set of sensors allows determining the trajectory and orientation of the robot at any moment to complement the information from its navigation system. Initially, it was necessary to build controlled test environments that allow observing and getting to know the trajectory and the dimensions thereof for capturing the raw data of the mobile robotic platform. Subsequently, the data was processed off-line implementing a Kalman filter [18] to remove Gaussian noise from the signal; then the sensor fusion algorithm was implemented allowing estimating the trajectory and the absolute orientation of the mobile robotic platform. It is worth mentioning that all the hardware elements used in the development of this idea are inexpensive, thus allowing the necessary testing, the data analysis and the interpretation of the outcome to be replicable and to be used as an educational tool within courses of undergraduate and master level. Finally, this initiative aims to explore the link between mobile robotics and Inertial Measurement Units to determine the trajectory and the absolute orientation as well as encouraging future studies where this system can be used to detect errors of other typical navigation systems, e.g. to determine the required power according to the inclination of the surface and to detect collisions, among many other applications.

2. Methodology

2.1. IMU MPU9150

The MPU-9150 is an integration of three sensors: 3-axis gyroscope, 3-axis accelerometer and 3-axis magnetometer. The gyroscope has a operating ranges of ±250, ±500, ±100 and ±2000 dps. The accelerometer has a programmable range of ±2, ±4, ±8 and ±16 g. The power range of the circuit is 2.4v - 3.46 v.

2.2. 1D Kalman Filter

The output signal of the accelerometer has a high noise component that can be characterized as Gaussian noise. Therefore, it was necessary to implement a 1D Kalman filter. The equations are shown in Equation 1 where $\hat{X}_k$ is the estimated value, Z_k is the measured value and K_k is the Kalman gain.

$$\hat{X}_k = K_k Z_k + (1 - K_k)\hat{X}_{k-1} \tag{1}$$

To calculate K_k it is necessary to know or to estimate the error covariance of the process P, the error covariance of the observation R that in the case of the accelerometers can be obtained with the probability density function of the data while the accelerometer is static. If the value of R is known, the process shown in Equation 2 will be used, where m is a vector of previously acquired raw samples to obtain the nature of the noise. Note that in this case, since the application of the filter in a one-dimensional signal is scaled, the matrixA = 1, and the deviation Q should be adjusted to the speed of the system.

$$
\begin{aligned}
&K_k = \frac{P_k}{P_k + R} \\
&\hat{X}_k = \hat{X}_{k-1} + K_k(Z_k - \hat{X}_{k-1}) \\
&P_k = (1 - K_k)\, P_{k-1} \\
&Where, X_{(0)} = mean(m); P_{(0)} = mean(m); R = \sigma^2(m)
\end{aligned}
\tag{2}
$$

2.3. Orientation and magnitude

After applying the calibration procedures [18], and implement 1D Kalman filter, model described in Equation 3 can be applied to estimate the trajectory of the mobile platform. Importantly calibration algorithms are essential to the correct functioning of this model [18].

$$
\begin{aligned}
Orientation &= tan^{-1}\left(\frac{mc_y}{mc_x}\right) \\
Magnitude &= \sqrt{ac_x^2 + ac_y^2}
\end{aligned}
\tag{3}
$$

2.4. Reference system test

To develop the model, it is necessary to have a reference system. For this purpose we used a line follower, which moves in geometric figures. The figures were printed on white paper with black ink. Twelve basic geometric figures were taken, three squares, three circles, three rectangles and three ovals. The dimensions of the squares are, 15 cm, 25 cm and 35 cm respectively, those of the circles are 13 cm, 23 cm and 33 cm of radius; for the rectangle the dimensions are 36 cm 25 cm, 35 cm 26 cm and 20 cm 31 cm, and finally the ovals dimensions are 49 cm 33 cm, 34.5 cm 23 cm and 18.5 cm 13 cm. By using a line follower robot, a route of three full circulations was performed of each figure, and data was collected with an Inertial Measurement Unit (IMU), reference IMU9150, which records 9 degrees of freedom. This data was used as training data and validation.

3. Results

To apply the linear Kalman filter, it must be absolutely assured that the nature of the noise is Gaussian. Figure 1 shows the PDF of the static signal of IMU9150 on the x-, y- and z-axis (accelerometer) before and after applying the Kalman filter. These results were obtained by placing the accelerometer on an anti-vibration table with a sampling frequency of 150 Hz.

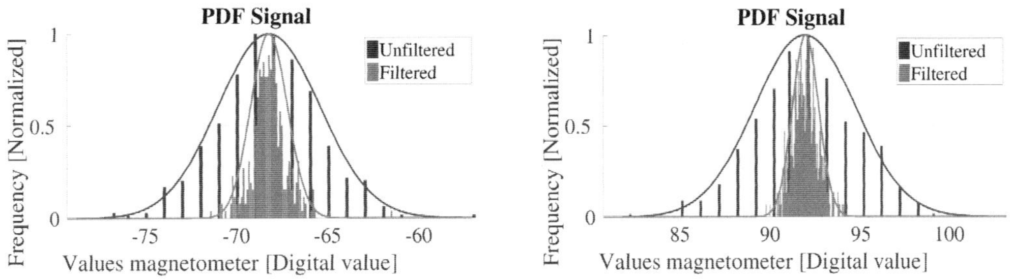

(a) (b)

Figure 1. (a) Kalman filter m_x (b) Kalman filter m_y

IOP Conf. Series: Materials Science and Engineering **138** (2016) 012005 doi:10.1088/1757-899X/138/1/012005

Figure 2 shows the evolution of orientation and reconstruction of the trajectory using the oval. As shown, four cycles were completed. Accumulation of bias is the difference in paths.

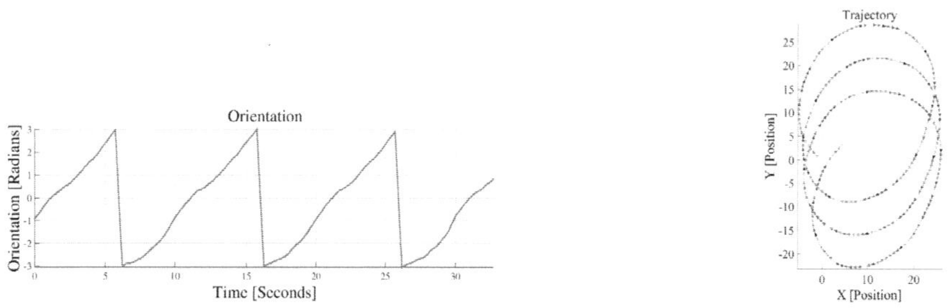

(a) (b)

Figure 2. (a) Oval orientation (b) Oval path

Figure 3 shows the rectangular tracking path. It is clear that the corners are not defined, but this effect is in the way the mobile platform monitors the line in the reference system.

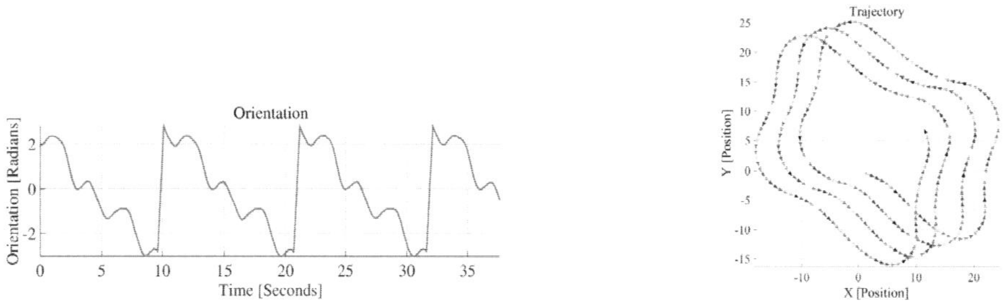

(a) (b)

Figure 3. (a) Rectangle orientation (b) Rectangle path

Finally in Figure 4, shown the path tracking by dynamic bias correction adjusting it with the temperature value. Clearly, the center of the road moves less than in previous experiments.

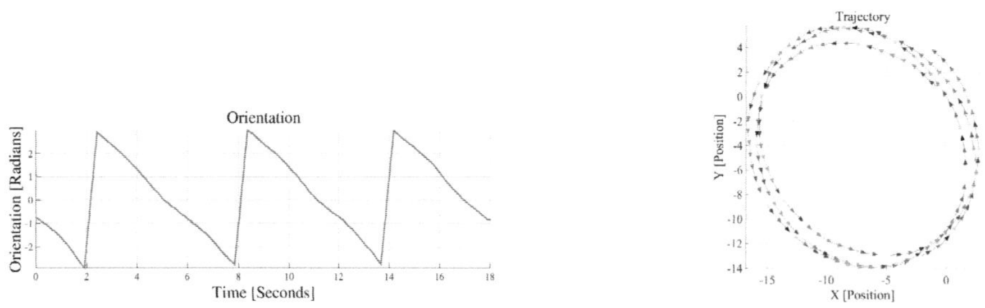

(a) (b)

Figure 4. (a) Circle orientation (b) Circle path

4. Conclusions

In this paper a model to approximate the path of a mobile robot is presented using data from a magnetometer and a gyroscope; and further to approximate the size of the path with the values of the accelerometer. This work is presented as a first approximation to a model that allows for absolute tracking of a mobile robot. Testing the implementation of a 1D Kalman filter was performed for each of the three accelerometer measures and their effectiveness was shown. Importantly, the models effectiveness depends heavily on the quality of the measured signal. Finally, the approach shown in this article allows demonstrating the utility of the Inertial Measurement Units for kinematic models. Advances in this area are adaptable to the applications mentioned above such as sports, the analysis of human activity and the treatment and diagnosis of diseases.

5. Acknowledgments

This work is part of the project "Development of an inertial measurement to obtain and record biomechanical variables in athletes" with ID 1102-626-38784, of the Automática y Electrónica Group COL0053581 attached to the Instituto Tecnológico Metropolitano and co-financed by the Liga de Natación de Antioquia and Fundación Universitaria Luis Amigó.

References

[1] Paul T, Singh J, Nayak M M, Rajanna K and Kumar M S 2008 *International Journal on Smart Sensing and Intelligent Systems* **1** 1019–1030 URL http://www.s2is.org/Issues/v1/n4/papers/paper11.pdf
[2] Banos O, Toth M A, Damas M, Pomares H and Rojas I 2014 *Sensors (Basel, Switzerland)* **14** 9995–10023 ISSN 1424-8220 URL http://www.mdpi.com/1424-8220/14/6/9995/htm
[3] Sampaio Santana D D, Massatoshi Furukawa C and Maruyama N 2013 *Latin America Transactions, IEEE (Revista IEEE America Latina)* **11** 1015–1021 ISSN 1548-0992
[4] Bergamini E, Ligorio G, Summa A, Vannozzi G, Cappozzo A and Sabatini A 2014 *Sensors* **14** 18625–18649 ISSN 1424-8220 URL http://www.mdpi.com/1424-8220/14/10/18625/
[5] Boerema S T, van Velsen L, Schaake L, Tönis T M and Hermens H J 2014 *Sensors (Basel, Switzerland)* **14** 3188–206 ISSN 1424-8220 URL http://www.mdpi.com/1424-8220/14/2/3188/htm
[6] Seel T, Raisch J and Schauer T 2014 *Sensors (Basel, Switzerland)* **14** 6891–909 ISSN 1424-8220 URL http://www.mdpi.com/1424-8220/14/4/6891/htm
[7] Huang Y C, Chen T L, Chiu B C, Yi C W, Lin C W, Yeh Y J and Kuo L C 2012 *Parallel Processing Workshops (ICPPW), 2012 41st International Conference on* (IEEE) pp 505–513
[8] Slawson S, Bindel A and West A 2010 *Signal Processing* URL http://ieeexplore.ieee.org/xpls/abs_all.jsp?arnumber=5742547
[9] Aminian K, Trevisan C, Najafi B, Dejnabadi H, Frigo C, Pavan E, Telonio A, Cerati F, Marinoni E C, Robert P and Others 2004 *Gait & posture* **20** 102–107
[10] Howcroft J, Kofman J and Lemaire E D 2013 *J Neuroeng Rehabil* **10** 10–91
[11] Özdemir A T and Barshan B 2014 *Sensors (Basel, Switzerland)* **14** 10691–708 ISSN 1424-8220 URL http://www.mdpi.com/1424-8220/14/6/10691/htm
[12] Wendel J, Meister O, Schlaile C and Trommer G F 2006 *Aerospace Science and Technology* **10** 527–533
[13] Won S H P, Golnaraghi F and Melek W W 2009 *Industrial Electronics, IEEE Transactions on* **56** 1782–1792
[14] Yuan X, Yu S, Zhang S, Wang G and Liu S 2015 *Sensors* **15** 10872–10890
[15] Yuan K, Wang H and Zhang H 2011 *Computational Intelligence and Design (ISCID), 2011 Fourth International Symposium on* vol 1 pp 294–297
[16] Wan E, Van Der Merwe R and Others 2000 *Adaptive Systems for Signal Processing, Communications, and Control Symposium 2000. AS-SPCC. The IEEE 2000* (IEEE) pp 153–158
[17] Julier S J 2002 *American Control Conference, 2002. Proceedings of the 2002* vol 6 pp 4555–4559 vol.6 ISSN 0743-1619
[18] Botero V J S, Hernandez W and Fernandez E 2014 *International journal on Smart Sensing and iIntelligent systems* **7** 1631–1646 URL http://www.s2is.org/Issues/v7/n4/papers/paper10.pdf

IOP Conf. Series: Materials Science and Engineering **138** (2016) 012006 doi:10.1088/1757-899X/138/1/012006

An angle measurement system of high resolution for the upper limbs using a low-cost servomotor

J.-S. Botero V.[1], J.-P. Restrepo Z.[1], M.-T. De Ossa J.[2]

[1]Grupo AEyCC, Facultad de Ingenierías
[2]Grupo GI^{2}B, Facultad de Ciencias Exactas y Aplicadas.
Instituto Tecnológico Metropolitano ITM, Carrera 31 No. 54-10, Medellín, Colombia

E-mail: `juanbotero@itm.edu.co`

Abstract. In the here presented study, the biomechanical design and coupling of a servomotor as measuring element for determining the angle of elbow flexion in humans is presented. This task requires a digital servomotor with a 12-bit low charge encoder type "contactless absolute", which makes the holding torque negligible. Because the servomotor is used as a sensor and not as an actuator, and is expected to produce the least possible resistance to the movement of the elbow, this is a crucial point. Additionally, the biomechanical design of the structure for coupling the servomotor was carried out considering the different movements of the arm and forearm, and the necessity to not interfere with the natural movement of the arm. The measurement resolution allows obtaining the flexion angle to an accuracy of 0.088; and integrated into the embedded system used to communicate with the servomotor, that allows obtaining and analyzing data and temporarily integrating information for counting repeats or measuring the speed of movements, among others. This system will also be useful to calibrate and compare other compatible biomechanical analysis models, where the same movement is analyzed.

1. Introduction

The acquisition of position information and angular speed of the human elbow is very important and useful for the design and manufacture of prosthetics and orthoses, and can support the study of patients undergoing rehabilitation, saving time and manpower, because the patient does not need to go to a medical center and no specialist is needed to take measures. Also angular movements occurring in daily life or in sports can be studied. For this purpose, a technique that implements an electromechanical system based on an embedded servomotor with a high resolution, but a low-cost encoder, which is not used as an actuator but rather as an angular position sensor receiving biomechanical information, is used. This information is adequately transferred to a computer, where the end user can analyze and use it for any of its possible applications. These signals offer the possibility of man-machine interfaces controlled by natural signals from the human body. Furthermore, they offer the ability to properly intervene in processes, on occasion providing support to people with physical disabilities and in rehabilitation, or to develop electromechanical systems to assist or to enhance physical skills for military and industrial applications.

Biomechanical signals, such as the angular position of the joints of the human body, are used in various applications, e.g. rehabilitation [1], medical monitoring, entertainment, military industry, product design, etc. Understanding these signals is essential to search for appropriate techniques to acquire and implement these signals in systems and technical devices; as in this case, the measurement of the angular orientation of the human forearm with respect to the elbow, with which analysis can be performed to design orthoses and prosthetics, or to interact with virtual environments, etc. In similar cases to this technique like [2], where the absolute encoder AS5040 from Austria Microsystems with a resolution of 0.35and a PIC16f877A were used. Angular position measurements of the knee were taken and the systems reliability was analyzed using a goniometer and comparing with motion analysis records. For the processing of the information a Matlab based interface was used. In [3], where the movement of the human forearm is captured by a reliable sensing system, based, as in [2], on the magnetic linear encoder AS5040 evaluating a noninvasive technique to estimate the angular position by using it as a digital goniometer tracking the orientation of the human limb. Also, there are different techniques, seeking to detect, estimate or measure joint positioning, as in the case of [4], where kinematic models are used in robotics and methods to continuously assess the orientation angles of the shoulder and the elbow based on inertial units and Kalman filter. These measurements were compared with those obtained by an optical tracking system showing that the error was always less than 8. In [5] a method is evidenced that estimates the angle of orientation of the human elbow through the processing of electromyographic (EMG) signals obtained from the surface of biceps and triceps aimed at optimizing while minimizing the error of the measured angle with respect to the actual angle. In [6], with the implementation of gyroscopes and accelerometers, a method is applied to obtain the orientation of a joint in static or dynamic state, using also speed and angular acceleration, as well as a Kalman filter to obtain the final estimation of the orientation. Angular positioning information is important and useful in various studies and applications making it necessary to find appropriate techniques, which offer reliable and easy to use information; thus requiring an electromechanical system able to monitor the angular position of the forearm relative to the elbow without limiting in any way the free movement of the individual. This system is based on the MX-12W servomotor from ROBOTIS line called Dinamixel [7], and which is not used as an actuator, but as an angular position acquisition device with a resolution of 0.088, capable of analyzing smaller displacements than the encoder implemented in [2] and [3].

Reliable information of biomechanical signals is of great importance in the rehabilitation of patients who suffered strokes. As shown by the World Health Organization [8], those can cause apoplexy and other damages and should be treated quickly to achieve the best possible recovery. So, different techniques have been created to support this rehabilitation based on biomechanical signals. In [9] a robotic training device based on surface EMG, which actively promotes the resumption of muscle signals, is implemented for hand rehabilitation in patients suffering from stroke. This system can be adjusted to different finger sizes, thus fitting many patients. In [10] they use a game, controlled by EMG, to rehabilitate patients in a didactic manner; and in [11] a robotic exoskeleton is meant to help people with arm and forearm problems in their daily lives. This exoskeleton has one degree of freedom and the control supports the movement of the elbow by manipulating the angular position based on biological signals that reflect the intention of the movement. Other consequences of brain injuries, such as hemiparesis and hemiplegia are also treated with various techniques, where, as shown in [12], the evaluation of motor skills plays a critical role. In case of hemiplegia, devices as shown in this article can be used to specifically detect the motion decrease of the elbow. Also it serves as a reference to generate myoelectric signals-based control. In addition, models of muscle monitoring can be created, as in [13], where EMG is used to study the elbow in dynamic situations including position and angular velocity.

It is worth mentioning that all the hardware elements used in the development of this idea are inexpensive, thus allowing the necessary testing, the data analysis and the interpretation of the outcome being replicable and using it as an educational tool within courses of undergraduate and master level. Finally, the system developed and integrated in this work includes the adaptive biomechanical design, the prototyping and manufacturing, the coupling with the servomotor and the program for the embedded system that allows translating the flexion information into meaningful information for the biomechanical analysis. Such initiatives aim at integrating low-cost technologies in applications that enhance the acquisition of information in rehabilitation systems, sports or the monitoring of physical activity in general, and thus, can have an impact on the customers quality of life.

2. System Description

2.1. Servomotor

The MX-12W is a high precision servomotor with the ability to measure positioning, torque, internal temperature and engine speed. It will be used to detect the angular position in the same way as with a digital goniometer and will also be used to measure the speed. MX-12W was selected due to its hardware, since its weight and size is useful in a system that will be used for measuring biomechanical signals. Although it will be non-invasive method, the individual will have to wear the electromechanical system like a garment. It has a positioning resolution of 0.088, which enables high-precision measurements with an input range of 0 to 360 and an output of 12 bits. It has a span of 360, which is more than sufficient for the required measurement, given that the angular movement of the elbow is not more than 150 according to [14]. Its physical dimensions are 32mm x 50mm x 40mm and it weighs only 54.6 grams.

2.2. Electromechanical system and control

For the correct measurement of the angular position, five pieces were used that were designed with CAD software and molded using a DREMEL 3D printer from Robert Bosch Tool Corporation. These pieces give rigidity to the structure allowing the servomotor to rotate regarding the degree of freedom of the elbow and allowing the mechanism being adjusted for different sizes of arm and forearm. The pieces shown in Figure 1 (a) are linked through the servomotor and they will also be fixed with screws to two plaster casts that are located directly to the arm and forearm, respectively. Furthermore, the embedded system of the servomotor, the OPEN CM09.04, which processes the information, is used for receiving data. Figure 1 (b) show the complete electromechanical system, which has not yet been attached to the connecting elements of the arm and forearm. To control the servomotor a brief treatment of the signal, which is already integrated, is required.

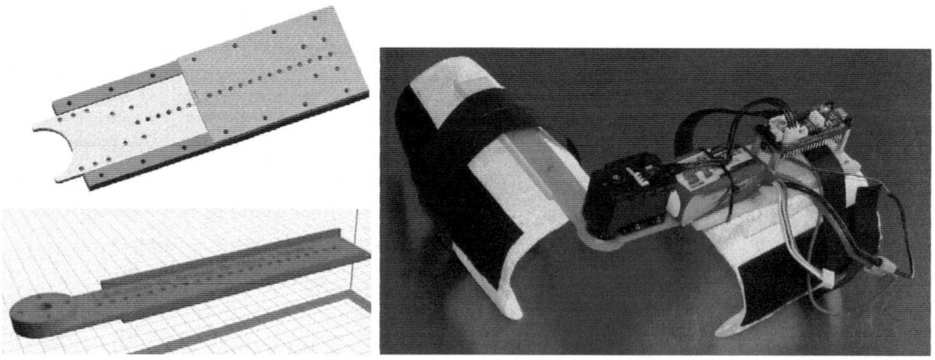

(a) (b)

Figure 1. (a) Arm piece and forearm piece (b) Arm piece with the embedded system

3. Results

3.1. Control

Figure 2 show the prototype system which will monitor the angular displacement. In this case it was set to be placed in a healthy person measuring 1.75 m in height.

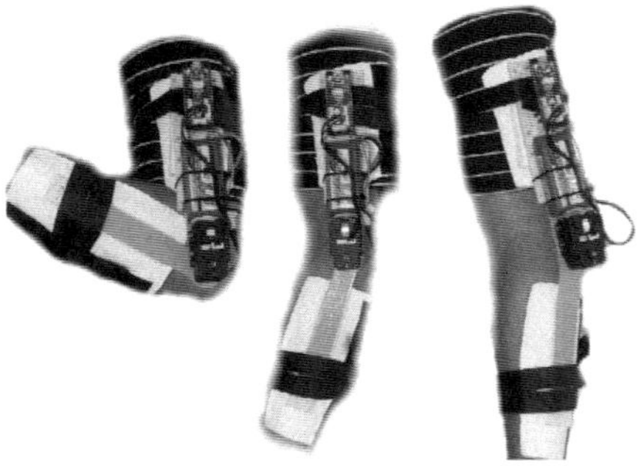

Figure 2. Prototype system

3.2. Data acquisition

Figure 3 shows the acquisition of the elbow movement in several repetitions and three speeds by the same person. An external LED was used to indicate the time each flexion took, that is to say 1 (Fast), 3 (Normal), and 4 (Slow) seconds, respectively. It was observed that the data acquired by the system allows deriving relevant information for biomechanical analysis, such as the number of repetitions per minute, the speed of the movement of flexion and extension, and the continuity in the movement. This data is useful in the development of physiotherapy aimed at rehabilitation or analysis of performance increase in spots activities, because it allows traceability in the time and in the development of the activity, and eliminates the need for supervision.

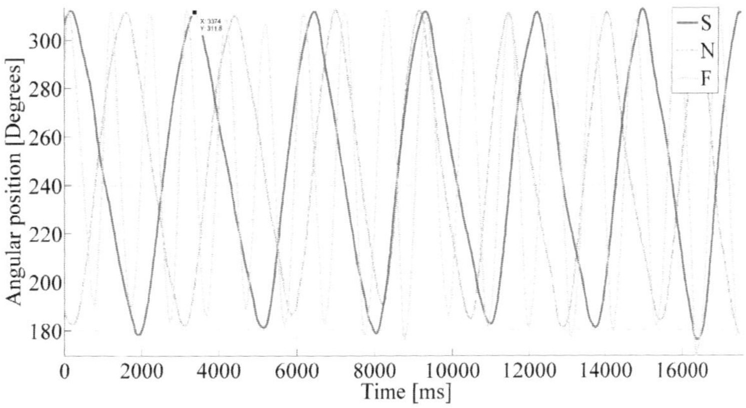

Figure 3. Acquired position for a test subject

A comparison of the angular position of arm and forearm in three different individuals is shown in Figure 4. It can be observed that, although the same external indicator was used to mark the moment of the movement, there are different response times, which is also an important issue for biomechanical analysis and reaction time measurements.

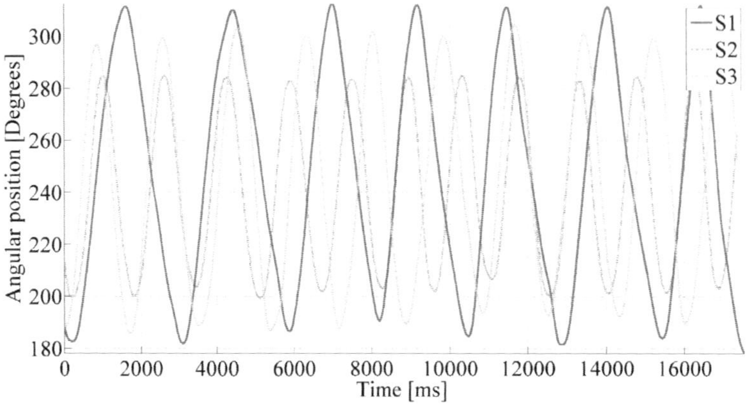

Figure 4. Acquired position for three test subject

4. Conclusion

This work is based on a measuring technique to monitor the angular position and speed of the human elbow using an electromechanical system, built with an intelligent, high resolution, but low-cost servomotor. Furthermore, the mechanical part of the device was fabricated with a 3D printer, taking the possibility of changing the size to suit different sizes of the arm and forearm into account; thus, offering the opportunity of conducting practical studies wherever data of the position or speed of the elbow is required for daily life tasks from people with lesions in this area, or disabilities, or to obtain reliable reference data from healthy people. In addition to using the servomotor as a sensor, rather than as an actuator, the system has an integrated OpenCM9.04 board for data processing and sending . Future work aims at the improvement of the biomechanical system to measure other joints, bearing in mind that implementation in the knees is also possible

The here presented application is aimed at offering an accurate measuring system and traceability of an exercise that is commonly used to determine the physical condition of patients in rehabilitation or the progress in muscle response in athletes. The main objective is that by developing a low-cost system, where the mechanical parts become public domain, it will be possible to offer alternative measurement methods to persons or entities that do not have access, for cost reasons, to traditional systems; and besides, to allow direct supervision by physiotherapists or coaches during the exercise.

5. Acknowledgments

This work is part of the project "Development of an inertial measurement to obtain and record biomechanical variables in athletes" with ID 1150-626-38784, of the Automática y Electrónica Group COL0053581 attached to the Instituto Tecnológico Metropolitano and co-financed by the Liga de Natación de Antioquia and Fundación Universitaria Luis Amigó.

References

[1] Ayala Lozano J F, Urriolagoitia Sosa G, Romero Ángeles B, Torres San-Miguel C R, Aguilar Pérez L A and Urriolagoitia Calderón G M 2015 *Revista Colombiana de Biotecnología* **17** 79–90 ISSN 1909-8758

[2] Domínguez G, Cardiel E, Arias S and Rogeli P 2013 *2013 World Congress on Nature and Biologically Inspired Computing, NaBIC 2013* (Fargo, ND) pp 1–4 ISBN 9781479914159

[3] Zimoga M A and Padmanaban L 2014 *International Conference on Information Communication and Embedded Systems (ICICES2014)* 978 (Chennai: IEEE) pp 1–4

[4] El-Gohary M and McNames J 2012 *IEEE Transactions on Biomedical Engineering* **59** 2635–2641 ISSN 0018-9294

[5] Yu H J, Lee A and Choi Y 2011 *IET Signal Processing* **5** 767
[6] Petruska A J and Meek S G 2011 *2011 IEEE/RSJ International Conference on Intelligent Robots and Systems* (San Francisco, CA: IEEE) pp 3632–3637 ISBN 978-1-61284-454-1 ISSN 2153-0858
[7] ROBOTIS 2010 MX 12 e-Manual v1.25.00
[8] WHO 2015 Stroke, Cerebrovascular accident
[9] Ho N S K, Tong K Y, Hu X L, Fung K L, Wei X J, Rong W and Susanto E a 2011 *2011 IEEE International Conference on Rehabilitation Robotics* vol 2011 (IEEE) pp 1–5 ISBN 978-1-4244-9862-8
[10] Xing S and Zhang X 2010 *2010 IEEE Conference on Robotics, Automation and Mechatronics* (Singapore: IEEE) pp 32–36
[11] Kiguchi K, Kariya S, Watanabe K, Izumi K and Fukuda T 2001 *IEEE Transactions on Systems, Man and Cybernetics, Part B (Cybernetics)* **31** 353–361 ISSN 10834419
[12] Zhang M, Lange B, Chien-Yen Chang, Sawchuk A a and Rizzo A a 2012 *2012 Annual International Conference of the IEEE Engineering in Medicine and Biology Society* (San Diego, CA: IEEE) pp 6111–6115
[13] Song R and Tong K Y 2005 *Medical & Biological Engineering & Computing* **43** 473–480 ISSN 0140-0118
[14] Montenegro M, Lugo A, Nunez V, Gonzalez K and Zaldivar U 2013 *XV COMROB 2013, Facultad de Electromecánica, Universidad de Colima* 1–6

Development of an aerial counting system in oil palm plantations

Jhany Zulyma Miserque Castillo[1], Rubbermaid Laverde Diaz[1] and Claudia Leonor Rueda Guzmán[2]

[1]Student, Universidad Pontificia Bolivariana, Autopista Piedecuesta Km. 7, Santander, Colombia
[2]Associated Professor, Universidad Pontificia Bolivariana, Autopista Piedecuesta Km. 7, Santander, Colombia

E-mail: jhanymiserque@gmail.com, rubbermaidlaverde@gmail.com, claudia.rueda@upb.edu.co

Abstract. This paper proposes the development of a counting aerial system capable of capturing, process and analyzing images of an oil palm plantation to register the number of cultivated palms. It begins with a study of the available UAV technologies to define the most appropriate model according to the project needs. As result, a DJI Phantom 2 Vision+ is used to capture pictures that are processed by a photogrammetry software to create orthomosaics from the areas of interest, which are handled by the developed software to calculate the number of palms contained in them. The implemented algorithm uses a sliding window technique in image pyramids to generate candidate windows, an LBP descriptor to model the texture of the picture, a logistic regression model to classify the windows and a non-maximum suppression algorithm to refine the decision. The system was tested in different images than the ones used for training and for establishing the set point. As result, the system showed a 95.34% detection rate with a 97.83% precision in mature palms and a 79.26% detection rate with a 97.53% precision in young palms giving an F1 score of 0.97 for mature palms and 0.87 for the small ones. The results are satisfactory getting the census and high-quality images from which is possible to get more information from the area of interest. All this, achieved through a low-cost system capable of work even in cloudy conditions.

1. Introduction

Colombia is the largest producer of palm oil in America and fourth in the world; thus the oil palm cultivation is a significant economic activity. These plantations cover vast areas affected by natural phenomena as plagues, diseases, storms and streaks of lightning [1]. These cause that the number of palms changes quickly and every plant represents money, time, and services. So, it needs to perform a census that can frequently be updated with high precision, so that provides as much as possible information about the content of the plantation allowing the farmers to have more control of the required work and the physical and economic resources of the company.

For a long time, this process has been done with workers going through the entire growing counting the number of palms and writing this information with pencil and paper. This technique takes too much time and so is not performed often, besides it is subject to human error. Currently, some companies have turned to modern technologies such as satellites and radars to get aerial views of the plantations and new information for the monitoring [2][3][4]. But, the typical cloud from the tropical weather where the oil

palm grows makes the task even more challenging, and the access to these resources has a high price. In the agricultural industry, the UAVs have been used to spread chemicals in the cultivations [5], and the obtained images have been used to perform remote monitoring of the products [6] and to create topographic maps with information about the ground conditions [2].
Meanwhile, the image processing and the machine learning applies in the assessment of the ripeness stage of fruits such as tomatoes [7], avocados [8] and the fruit of the oil palms [9]. A new use for this technology has been recently proposed: to identify plants that present specific colors which are characteristic of some diseases in the oil palms [10] and despite, this has been tested in domestic plants, its performance has never been verified in real oil palm plantations.
This paper presents the development of a system that uses a UAV to fly over the plantation and to get images that are processed with a photogrammetry software to create orthomosaics of the interest zones and then obtains an automatic census of the palms using a detection and counting software. Its general structure is described in figure 1 as a block diagram.

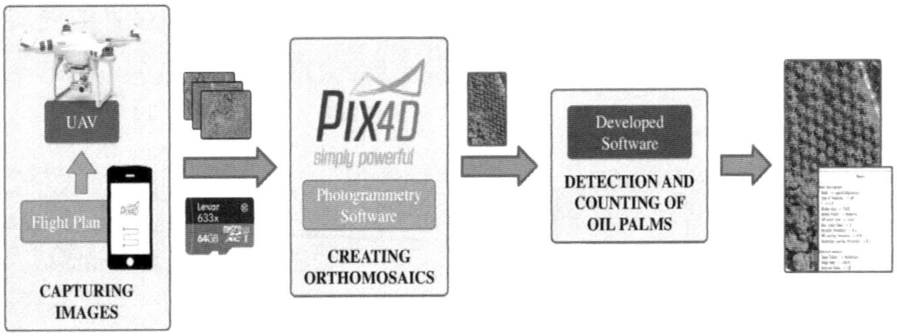

Figure 1. Block diagram of the general structure of the system.

2. Capturing images

Data is obtained using a UAV DJI Phantom 2 Vision+ controlled by a mobile app installed on a device with an iOS operating system where is possible to set the area of the plantation that will be covered by the UAV and the points where photos must be captured.

2.1. Factors influencing the selection process of the UAV

To define an adequate UAV model, a comparative analysis is performed between some models available in the market considering the most important aspects according to the project needs such as the capacity of integration with a photographic camera, cost, camera type, flight time, compatibility with photogrammetry tools and autonomous flight. Besides, it is important to considerate that oil palm plantations are distributed by lots that usually are only accessible through a dirt road that subdivides them, and when the palms are mature the ramifications stick out over the road limiting space. So the vertical takeoff UAVs such as helicopters and multi-rotors are an adequate solution and so the selection process is restricted only to this type of aircraft.

2.2. Comparison of some available models

The preceding factors were considered in three UAV references available in the Universidad Pontificia Bolivariana (Bucaramanga) to determine if any of them fulfilled the project needs: AR.Drone (Parrot), AR.Drone 2.0 (Parrot) and Phantom 1 (DJI) equipped with a GoProHERO3 camera. The AR.Drone model was rejected because it was not capable of taking photos or implementing a GPS system[11]. Some tests were performed with the other two models to compare their behavior, and it concluded that the AR.Drone 2.0 did not have the required stability to carry out long flights in an open field, meanwhile the Phantom 1, despite not having this problem, did not offer the possibility to execute autonomous flights.

2.3. Model selection

The Phantom 2 Vision+ (DJI) was an adequate UAV for the task because it integrated all the advantages of the analyzed models. This UAV allows realizing long and stable autonomous flights to get high-quality images since it implements an electronic gimbal mount for the camera with stabilization in the three axes.

3. Creating orthomosaics

Figure 2 shows a group of photos acquired by the UAV in an area of the plantation. It shows the perspective distortion problem generated by the camera on-board the UAV due to its wide-angle lens which offers a wide visual field but causes distortions in the image. This is the reason why the palms are recognizable in the central zone of the photo, but as it reaches the borders, the perspective distortion increases making harder to differentiate between palms.
An orthomosaics allows getting, from a succession of pictures, an image as the presented in figure three which covers vast extensions and permit to observe all the elements under the same orthogonal projection without perspective distortion. For this, it is necessary to use photogrammetry tools [12] as the mobile app Pix4Dmapper Capture utilized to program a flight path as a grid that guarantees that all the photos are taken at the same height and with a constant overlap percentage. Later, these photos were processed in a computer with the software Pix4Dmapper to create the corresponding orthomosaics.

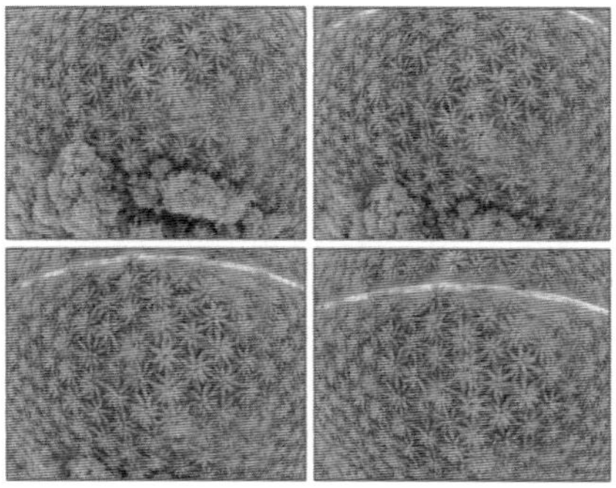

Figure 2. Photos from wide-angle lens with the perspective distortion problem.

Figure 3. Orthomosaic with constant orthogonal projection for all the elements.

4. Detection and counting software

4.1. Training of the logistic regression model

To detect the palms, the system implements a supervised machine learning [13][14][15] model that requires an image database labeled with the corresponding categories. First, a person selects the positive (palms) and negative (backgrounds) samples in the four orthomosaics used for training using an interface developed in Python. As result, a database with 450 positives samples and 2500 negatives samples are obtained getting a total of 2950 samples as the ones presented in figure 4. Every sample has the size of a window of 72x72 pixels in which is possible to frame the center of a grown palm without including all its branches. Each one is described using the Uniform-LBP operator (Local Binary Pattern) so that each pixel gets a value according to the texture patron that surround it. To make the model more robust, the window is divided into several overlapping blocks, so the U-LBP blocks are set with a size of 12x12 pixels with an overlap of 6 pixels between them, obtaining a total of 121 LBP blocks for a 72x72 window. Each U-LBP [16] block is represented by its histogram, which has 59 possible patrons, and the entire window is characterized by the sequence of these histograms. Thus, the characteristics vector of a window has 7139 values.

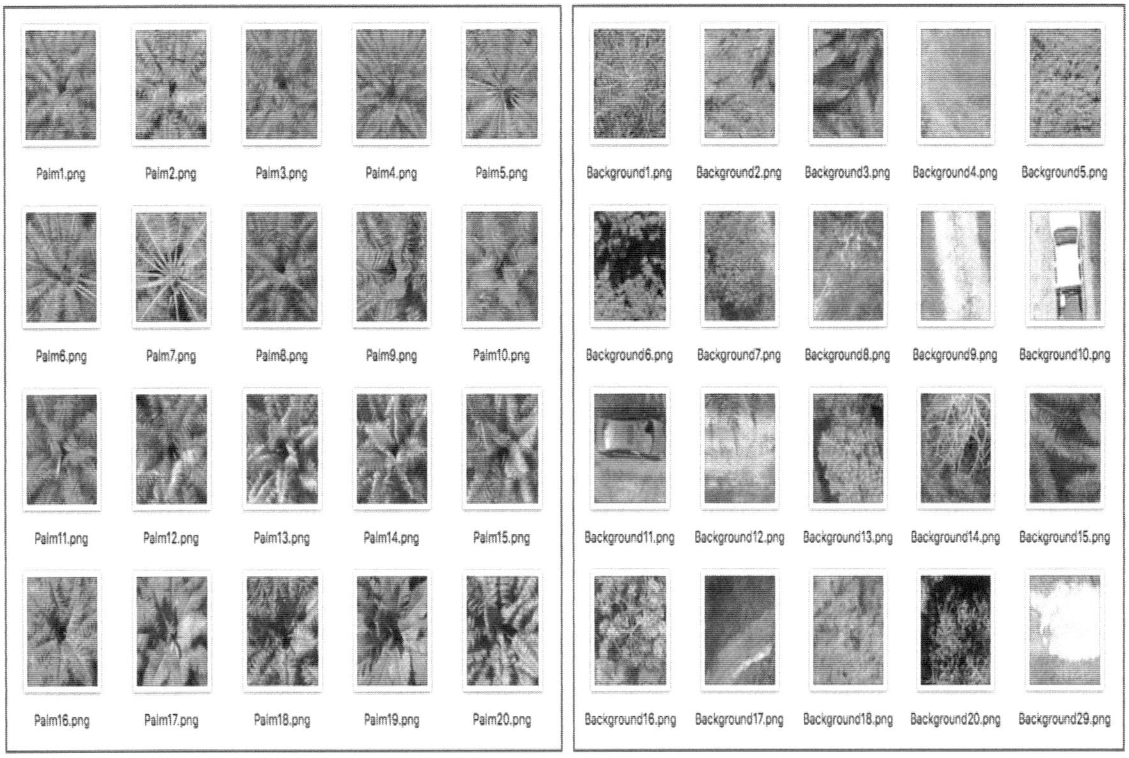

Figure 4. Examples of positive and negative samples from training database.

These samples are organized to train a logistic regression model [17] which gets a score of 1.0 when evaluates all the training samples, which means that is capable of classifying all the samples used for its training.

4.2. Software for detection and counting of oil palms

The block diagram in figure 5 presents the process to detect and count palms in an orthomosaic.

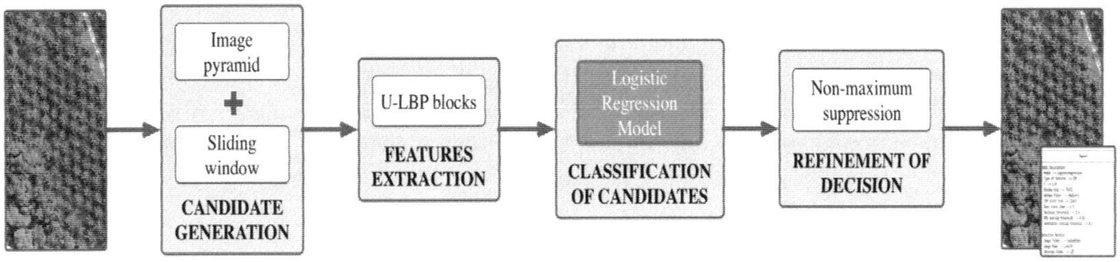

Figure 5. Block diagram of the software to detect and count palms.

First, the software converts the image to grayscale and identifies all the possible 72x72 windows where a palm can be located. Then, a Gaussian pyramid [16] is implemented to identify palms of different sizes in the various levels of the pyramid. But, only the first six levels of the pyramid are considered due to the limitation in the size that a palm can occupy in the image. Each level is scored with a sliding window, and each candidate window goes through the same U-LBP characteristics extraction process that the

training windows and thus feature vector with 7139 values describes the texture of the candidate window and allows the model to understand the type of contained object.
The model gives a score for every analyzed window indicating the probability to contain a palm. When this score exceeds the established decision threshold (0.5), the window qualifies as palm. But, one palm can be detected by several boxes that obtained enough score to surpass the threshold and hence the decision is refined with the non-maximum suppression technique [18]. The length of the vector with the last detection boxes indicates the number of detected palms, which is presented in a report. Besides, the bounding boxes are drawn in the images as well as the scores given by the model.

5. Evaluation

The performance of the system is evaluated in new orthomosaics, different to the ones used to train and adjust the system: one with mature palms and one with small plants.

5.1. Evaluation metrics

A per-image evaluation technique is used. It requires knowing the correct location of the palms in the image, which is made by a human expert manual selection. Then, the detection software is run for the first orthomosaic without including the NMS algorithm, so all the possible detections are considered. The NMS is applied changing the threshold from -2 to 2 with 0.02 steps and the detections for every threshold are saved.
For each operation point, the detections are compared against the coordinates annotated for the human expert and an overlap of 30% or more indicates a correct detection. This allows calculating the False Positives per Image (FPPI) vs. Miss Rate, and Recall vs. Precision curves. Every point in this plots describes the performance of the system for a particular threshold (operation point). Besides, An F1 score is computed, which considers the detection rate and the precision.

5.2. Results

The results from the per-image evaluation [18] in the orthomosaic with mature palms are presented in the figures 6 and 7. For the chosen operation point (threshold = 0.5) which minimizes the miss rate and the FPPI, and maximizes the recall and precision, the system achieves a detection rate of 95.34% with a miss rate of 4.66% indicating that is capable of identifying most of the relevant results. The precision of 97.83% shows that the system presents more relevant results than irrelevant, so there are more true detections than false positives, which have an average of 5 FPPI. Considering that each orthomosaic includes between 150 to 250 palms, among 7 to 12 palms are not identified by the system, but the average of FPPI counteracts this value. Finally, for this case, the system obtains an F1 score of 0.97.
When the per-image evaluation is implemented in the orthomosaic with small young palms, the performance graphics presented in figures 8 and 9 are obtained. The chosen operation point (threshold = -1.5) for this situation must be smaller than the previous section. The system achieves a detection rate of 79.26% with a miss rate of 20.74% decreasing the performance when the palms are small. The high precision of 97.53% shows that despite all the desired plants are considered, the detected are mostly true detections. Finally, for this type of palms, the system gets an F1 score of 0.87.

6. Conclusions

The high-quality aerial views obtained in the orthomosaics using the photogrammetry tools from Pix4Dmapper solved the perspective distortion problem. Moreover, it was possible to prove that the UAVs are a successful tool to monitor plantations hence they allow obtaining useful information cheaper than the satellites and without the cloud problems. Besides, they prevent the worker to go through the entire plantation facilitating the work conditions and updating the information with more frequency.

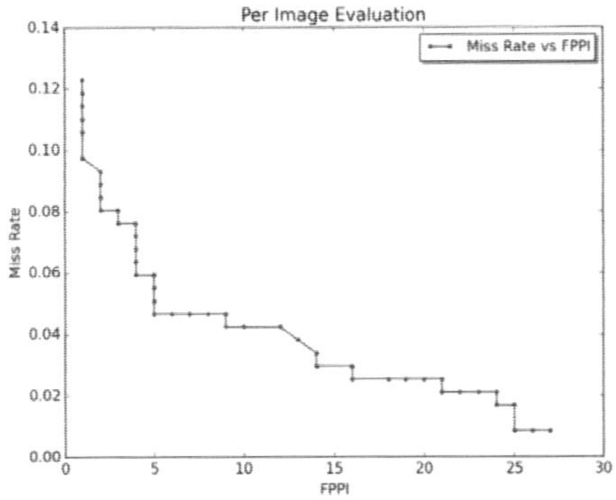

Figure 6. FPPI vs Miss Rate in mature palms

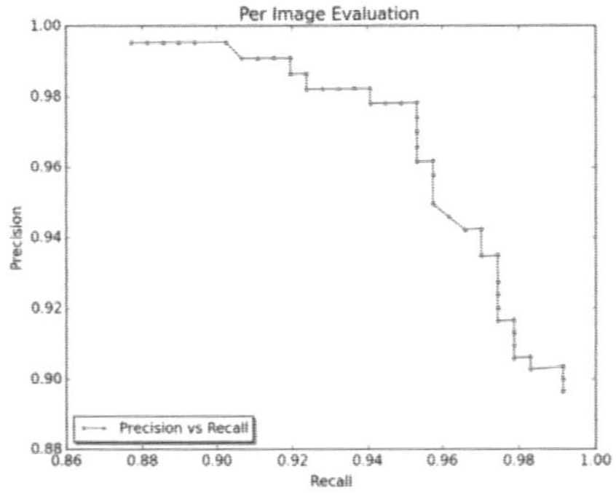

Figure 7. Recall vs Precision in mature palms

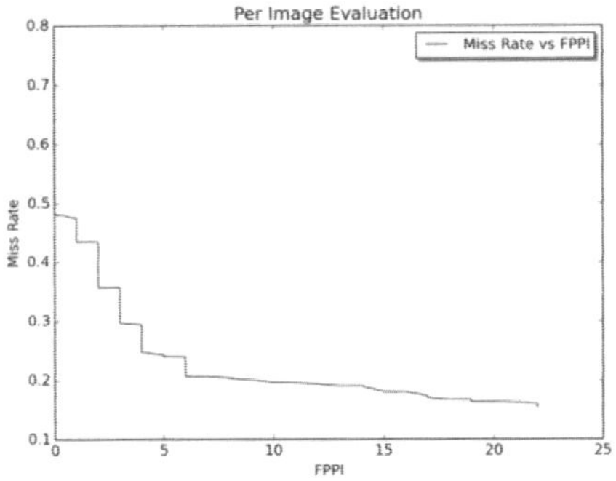

Figure 8. FPPI vs Miss Rate in young palms

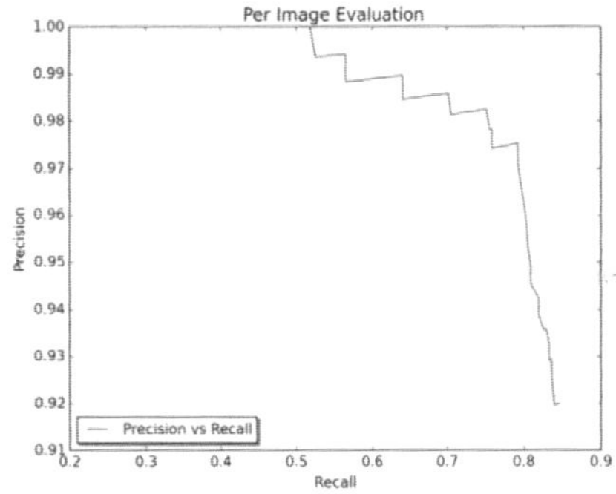

Figure 9. Recall vs Precision in young palms

7. References

[1] Chinchilla C 1998 *Palma.* **19** 242-6
[2] Pohl C 2014 *IOP Conf. Ser: Earth Environ. Sci.* **20** 012012
[3] Kwesi N 2012 *Oil palm mapping using Support Vector Machine with LANDSAT ETM+ data* (Enschede: University of Twente)
[4] Ng P 2013 *Palmas.* **34** 259-79
[5] Costa F, Ueyama J, Braun T, Pessin G, Osorio F and Vargas P 2012 *Geosci. Remote. Sens. Symp. (Munich) IEEE Int.* 5045-48
[6] Berni J, Zarco-Tejada P, Suarez L and Fereres E 2009 *IEEE Trans. Geosci. Remote. Sens.* **47**
[7] Zhang L and McCarthy M 2012 *Postharvest Biol. Technol.* **67** 37-43
[8] Guerrero E and Benavides G 2014 *IEEE Colomb. Conf. Commun. Comput. (Bogotá)* p.1-6
[9] Bensaeed O, Shariff A, Mahmud A, Shafri H and Alfatni M 2014 *IOP Conf. Ser: Earth. Environ. Sci.* **20** 012017
[10] Gómez S and Segura F 2012 *An aerial monitoring system (AMS) for detecting bud-rot diseases on oil palms* (Bogotá: Universidad de los Andes)
[11] Rincón V 2015 *Palmas.* **36** 25-41
[12] Diaz C 2013 *Adquisición de imágenes de bajo costo aplicadas a la agricultura de precisión usando vehículos aéreos no tripulados* (Quito: Universidad San Francisco)
[13] Neupane R, Sharma K and Thapa G 2002 *Agricult. Syst.* **72** 177-96
[14] Ohlmacher G and Davis J 2003 *Engin. Geology.* **69** 331–43
[15] Stephenson C, MacKenzie M, Edwards C and Travis J 2006 *Ecological Modelling* **193** 747- 58
[16] Pedregosa F, Varoquaux G and Gramfort A 2011 *The Jour. of Mach. Lear. Reser.* **12**, 2825-30
[17] Neubeck A and Gool 2006 *Pattern Recognition, 18th Inter. Confe. IEEE* **3** 850-855
[18] Bradley A 1997 *Pattern Recognition.* **30** 1145-59

IOP Conf. Series: Materials Science and Engineering **138** (2016) 012008 doi:10.1088/1757-899X/138/1/012008

Parametric uncertain identification of a robotic system

L Angel[1], J Viola[1] and C Hernández[1]

[1] Universidad Pontificia Bolivariana
Autopista Km7 via a Piedecuesta, Bucaramanga, Colombia

Email: (luis.angel, jairo.viola, carlos.hernandez)@upb.edu.co

Abstract. This paper presents the parametric uncertainties identification of a robotic system of one degree of freedom. A MSC-ADAMS / MATLAB co-simulation model was built to simulate the uncertainties that affect the robotic system. For a desired trajectory, a set of dynamic models of the system was identified in presence of variations in the mass, length and friction of the system employing least squares method. Using the input-output linearization technique a linearized model plant was defined. Finally, the maximum multiplicative uncertainty of the system was modelled giving the controller desired design conditions to achieve a robust stability and performance of the closed loop system.

1. Introduction

Usually, a parametric identification process is performed for a system under ideal conditions. That is, it does not take into account the presence of parametric uncertainness and external perturbations that alter the system dynamics. For a robotic system, parametric uncertain and external perturbations identification presents a challenge because these systems are non-linear, multivariable and exhibit a high degree of interaction between its links [2]-[4]. This paper proposes a parametric uncertainness identification for a robotic system of one degree of freedom (DOF). A MSC-ADAMS/MATLAB co-simulation model is built to simulate the robotic system dynamics. Then, the non-linear dynamic model parametric identification of the robot is performed using the recursive least square method [1]. After this, a linearized dynamic model of the robotic system is obtained applying the input-output linearization technique. This linear model belongs to the nominal plant and it represents a model of the system without considering external disturbances and parametric uncertainness. Then, the family of plants for the robotic system is calculated varying the mass, width and friction in the co-simulation model of the robotic system, applying the linearization technique for each condition. From the family of plants, the multiplicative uncertainty is computed for each member of the family and the maximum multiplicative uncertain profile of the system is determined. This maximum uncertain profile allows to find the desired design conditions for the controller which ensure the stability and robust performance of the robotic system [5]. The paper is structured as follows. First, a MSC-ADAMS/MATLAB co-simulation model is presented. Second, the parametric identification using recursive least square method is performed. Third, a linearized model of the robotic system is obtained by input-output linearization. Fourth, the family of plants is determined to obtain the maximum multiplicative uncertain profile and find the desired design controller specifications to reach a robust stability and performance.

2. ADAMS-MATLAB co-simulation dynamic model

In order to study the real-time dynamic behaviour of the robotic system is necessary to build a simulation model able to emulate all the dynamic actions applied over the robotic system during its operation, such as friction, Coriolis forces, gravity effect, and more. For this, MSC-ADAMS software was used. The robotic articulation of one DOF can be represented with a simple pendulum that receives an applied torque in order to generate the movement. The pendulum is placed in the same direction as the gravity, which must be compensated later for the development of the controller. The robotic system rotates around Z-axis, and the applied torque is perpendicular to the movement axis. Then, the MSC-ADAMS model is exported to MATLAB to modelling the parametric uncertain of the robotic system [6]. Figure 1a shows the MSC-ADAMS model of the robotic system.

3. Identification of the robotic-joint dynamic model

According to [1], the dynamic model for a robotic joint of 1DOF is given by (1)

$$\tau = [I_{r1} + m_1 {l_{cl}}^2]\ddot{q} + b_1\dot{q} + f_{c1}sign(\dot{q}) + [m_1 g l_{cl}]sin(q) \qquad (1)$$

where, τ is the applied torque, m_1 is the mass of the pendulum, l_{cl} is the gravity centre, g is the gravity force, $q, \dot{q}$ and $\ddot{q}$ are the position, speed and acceleration of the pendulum respectively, and b_1 and f_{c1} represent the viscous and Coulomb friction. (1) can be expressed in parametric form as is defined in (2).

$$\tau = \theta_1\ddot{q} + \theta_2\dot{q} + \theta_3 sign(\dot{q}) + \theta_4 sin(q) \qquad (2)$$

where $\theta_1 = I_{r1} + m_1 {l_{cl}}^2$, $\theta_2 = b_1$, $\theta_3 = f_{c1}$, $\theta_4 = m_1 g l_{cl}$. The identification of the dynamic model constants $\theta_1, \theta_2, \theta_3$ and θ_4 is performed using the linear regression method based on the recursive least-squares (RLS) algorithm. Based on the co-simulation model and the applied torque defined in (3) the identification process is implemented using the experimental data from position, speed, acceleration and applied torque of the joint.

$$Z = [2000\cos(2t) + 2000sin(10t)] \text{ N-mm} \qquad (3)$$

Applying the recursive least-squares algorithm, the constants of the dynamic model proposed in (2) corresponds to θ_1=150.0136, θ_2= -41.69, θ_3=-4.69, θ_4=5274.7. Figure1b shows the co-simulation torque trajectory Z against to the RLS model with torque trajectory Z with a correlation of 99.7%.

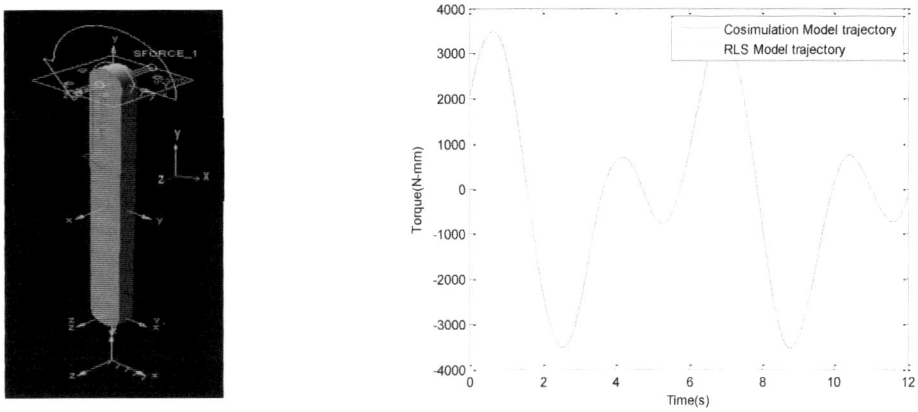

(a) (b)

Figure 1. (a) MSC-ADAMS model, (b) identification torque trajectory

4. Robotic system linearization using the input-output technique

To obtain the nominal plant and the family of plants, is necessary a linear model of the system. Since the robot dynamic is non-linear system as is described in (2), it requires a linearization technique to obtain its linear model. In this case, the input-output linearization technique was used [7]. The representation of the system in state variables is shown in (4), with $x_1 = q$ and $x_2 = \dot{q}$.

$$\begin{bmatrix} \dot{x_1}(t) \\ \dot{x_2}(t) \end{bmatrix} = \begin{bmatrix} x_2(t) \\ -\frac{\theta_2}{\theta_1}x_2(t) - \frac{\theta_3}{\theta_1}sign(x_2(t)) - \frac{\theta_4}{\theta_1}\sin(x_1(t)) \end{bmatrix} + \begin{bmatrix} 0 \\ \frac{1}{\theta_1} \end{bmatrix}\tau(t) \quad (4)$$

$$\text{If, } f(x) = \begin{bmatrix} x_2(t) \\ -\frac{\theta_2}{\theta_1}x_2(t) - \frac{\theta_3}{\theta_1}sign(x_2(t)) - \frac{\theta_4}{\theta_1}\sin(x_1(t)) \end{bmatrix} \text{ and } g(x) = \begin{bmatrix} 0 \\ \frac{1}{\theta_1} \end{bmatrix}$$

The system (4) can be expressed as:

$$\begin{aligned} \dot{x} &= f(x) + g(x)u \\ y &= h(x) = x_1 \end{aligned} \quad (5)$$

The linearization control law is given by (6)

$$u = \frac{1}{L_g L_{f(x)}}[-\beta_0 h_{(x)} - \beta_1 L_{f(x)} - L_{f(x)}{}^2 + v] \quad (6)$$

where v is the new entry of the system, β_0 and β_1 are the constants of the desired polynomial that represents the open-loop behavior of the linearized system. $L_{f(x)}, L_{f(x)}{}^2$ and $L_g L_{f\,(x)}$, are given by (7)-(10) using Lie's algebra.

$$l_f h(x) = \frac{\partial h(x)}{\partial x} f(x) = x_2 \quad (7)$$

$$l_g h(x) = \frac{\partial h(x)}{\partial x} g(x) = 0 \quad (8)$$

$$l_f{}^2 h(x) = \frac{\partial l_f h(x)}{\partial x} f(x) = -\frac{\theta_2}{\theta_1}x_2(t) - \frac{\theta_3}{\theta_1}sign(x_2(t)) - \frac{\theta_4}{\theta_1}\sin(x_1(t)) \quad (9)$$

$$l_g l_f h(x) = \frac{\partial l_f h(x)}{\partial x} g(x) = \frac{1}{\theta_1} \quad (10)$$

The desired characteristic polynomial for the linearized system is described in (11):

$$s^2 + \beta_1 s + \beta_0 = 0 \quad (11)$$

Parameters β_1, β_0, are computed so that the frequency response of linearized system ($P(s)$) be comparable to a second order linear model presented in (12).

$$P(s) = \frac{y(s)}{v(s)} = \frac{0.49}{1 + 1.30s + 0.496s^2} \quad (12)$$

5. Parametric uncertain identification and determination of robust operation conditions

Linearized model of the system presented in (12) correspond to the nominal plant. To obtain the family of plants, the robotic system is tested against parametric uncertainness in the width, mass and joint friction. Table 1 shows the original parameter of the robotic system and its percentage of variation due to parametric uncertainness described above.

Table 1 Range of variation of parametric uncertainness

Parameter	Nominal Value	Perturbation (%)
Mass (M)	2kg	± 10
Width (W)	4 cm	± 5
Joint Friction (F)	0.3 N	± 15

From each condition presented in Table 1, co-simulation model is modified to obtain the parameters θ_1 to θ_4 of the dynamic model (2) against each set of perturbations as show in Table 2. The + and – signs in Table 2 indicates the maximum and minimum values of the parameters mass (M), width (W) and joint friction (F). As shown in Table 2, the parameters of the dynamic model (2) varying according to the uncertain set presented. This indicates that the robotic system dynamics is different for each uncertain set applied. From the dynamic model parameters (2), the robotic system is linearized as a linear second order system descripted in (13) with ζ, T_w and k_p calculated for each condition as shown in Table 3. Frequency response of the family of plants is presented in Figure 2.

$$\widetilde{P_i}(s) = \frac{k_p}{1 + 2\zeta T_w s + (T_w s)^2} \tag{13}$$

Table 2. Robotic system dynamic model parameters variation for each parametric uncertainness.

Parameters	Nominal	F M W - - -	F M W + - -	F M W + - +	F M W - - +	F M W + + -	F M W + + +	F M W - + -	F M W - + +
θ_1	150.0136	139.7957	160.0754	160.03	138.95	160.11	160.07	139.08	138.91
θ_2	-13.6251	-18.872	-18.6263	-17.98	-17.74	-18.56	-17.82	-18.33	-17.69
θ_3	4.9657	30.5345	-0.0071	2.35	22.41	0.13	2.07	23.5	22.09
θ_4	5275	4759	5786	5784	4725	5785	5785	4725	4725

Table 3 Linearized family of plants for the robotic system.

Parameters	Nominal	F M W - - -	F M W + - -	F M W + - +	F M W - - +	F M W + + -	F M W + + +	F M W - + -	F M W - + +
k_p	0.49	1.7	0.193	0.19	1.69	0.193	0.1911	1.71	1.69
Tw	0.7	1.26	0.365	0.36	1.27	0.365	0.368	1.25	1.27
ζ	0.92	0.35	0.66	0.67	0.355	0.666	0.673	0.3624	0.355

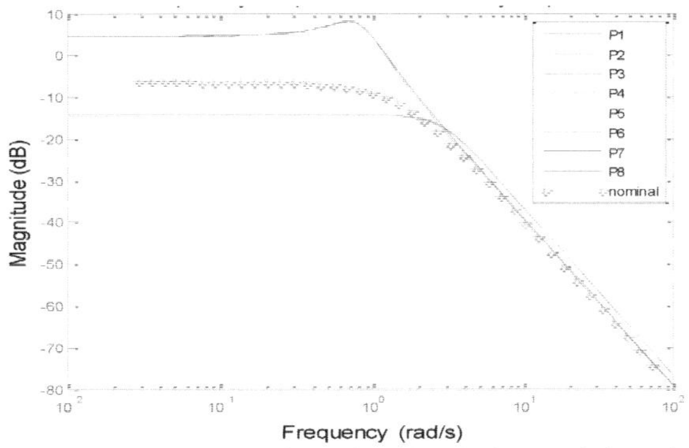

Figure 2. Frequency response of the family of plants of the robotic system

As shown in Figure 2, the nominal plant represented by * is placed between the variation range presented by the uncertainness descripted in Table 1. Once the family of plants is defined, the uncertain of each member of the family of plants is calculated using the multiplicative uncertain model [8] showed in (14).

$$|\delta P_i(s)| = \left|\frac{\tilde{P}_i(s)}{P_n(s)} - 1\right| \tag{14}$$

where $P_n(s)$ correspond to the nominal plant descripted by (12) and $\widetilde{P_i}(s)$ is the ith member of the family of plants. Frequency response of the multiplicative uncertain profile of each member of the family of plants is shown in Figure 3a. To find the conditions for robust performance and stability is necessary to determinate the maximum multiplicative uncertain profile. For this, the maximum uncertain profile is estimated according to the frequency response of the multiplicative uncertain for the family of plants shown in Figure 3. The maximum multiplicative uncertain profile reached is given by (15) and its frequency response is shown in Figure 3b.

$$\delta_{pmax}(s) = \frac{0.5s^2 + 5s + 8}{s^2 + 2s + 1} \tag{15}$$

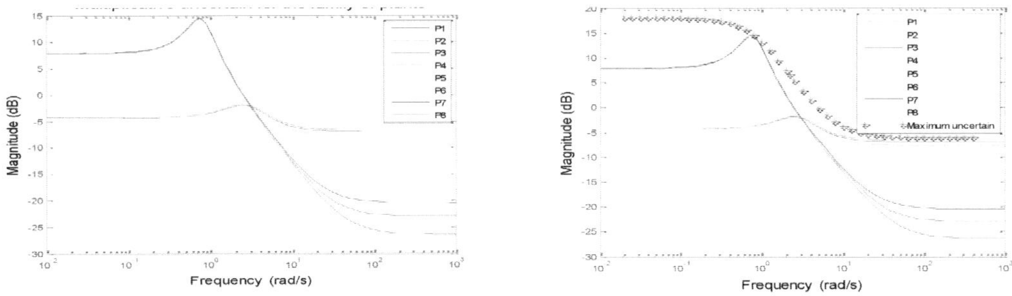

(a) (b)

Figure 3. (a) Multiplicative uncertain profile for the family of plants of the robotic system and (b) Maximum multiplicative uncertain profile of the family of plants.

As observed in Figure 3b, the maximum profile of multiplicative uncertain represented by * covers all the multiplicative uncertainness descripted by the family of plants. From [8], the closed loop desired specifications are calculated to achieve a robust stability and performance according to (16).

$$|T(s)| \leq \left|\frac{1}{\delta P_{max}(s)}\right| \tag{16}$$

where $|T(s)|$ is the magnitude of the complementary sensibility function and $|\delta P_{max}(s)|$ correspond to the magnitude of maximum profile of multiplicative uncertain. Figure 4 shows the function $|1/\delta P \max\ (s)\ |$ which restrict the closed loop bandwidth of the system to $0.7\ rad/s$. The bandwidth is related to the natural frequency $w_n$by (17).

$$w_b = w_n\sqrt{1 - 2\zeta^2 + \sqrt{(1 - 2\zeta^2)^2 + 1}} \tag{17}$$

Assuming $\zeta = 0.6$ which is equivalent to an 10% overshoot, from (16) $w_n = 0.61\ rad/s$, the desired closed loop function is (18). The frequency response of (18) is shown in Figure 5 which bandwidth is limited by the maximum inverse uncertain profile.

$$T(s) = \frac{0.037}{s^2 + 0.73s + 0.37} \tag{18}$$

The closed loop specifications to reach a robust stability and performance are given by (19) and (20) where (19) is the phase margin and (20) the gain crossover frequency. Solving (19) and (20), the design conditions for the controller are $pm = 60°$ and $w_c = 0.43\ rad/s$.

$$pm = 100\zeta \tag{19}$$

$$w_c = w_n\sqrt{\sqrt{4\zeta^4+1}-2\zeta^2} \tag{20}$$

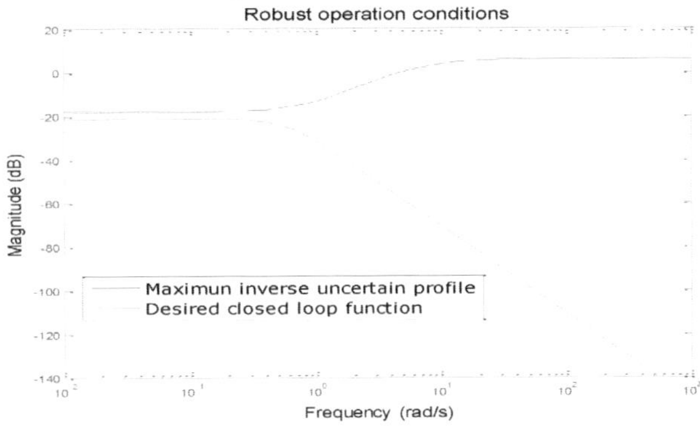

Figure 4. Desired robust operation conditions for the robotic system

6. Conclusions

This paper presented a methodology for the parametric uncertainness identification of a robotic system. A MSC-ADAMS/MATLAB co-simulation model was built for the parametric identification of the dynamic model using least square algorithm. The dynamic model was linearized using input-output linearization, obtaining the nominal plant. After that, a set of parametric uncertainness is defined to obtain the family of plants for the robotic system. Then the multiplicative uncertain of the family of plants is determined to find the maximum uncertain profile, which one establishes the closed loop system specifications to reach a robust stability and performance. Results show that when a robotic system is affected by external disturbances and parametric uncertainness, it results in many dynamical behaviours of the system, which must be covered for a controller to get a robust performance of the system. Employing the proposed methodology the desired design controller specifications to achieve a robust stability and performance against the presence of external disturbances a parametric uncertainness can be found. As well, this methodology may be applied to any non-linear system in order to achieve a robust stability and performance.

References

[1] Reyes Cortes F 2010 *MATLAB aplicado a Robotica y mecatronica* (Madrid, Alfaomega grupo editor) 460p
[2] Salloum R, Arvan M R and Moaveni B Identification 2013 uncertainty modelling, and robust controller design for an electromechanical actuator *International conference on Control, Automation, Robotics and Embedded Systems*
[3] Sadeghzadeh A 2014 Performance analysis for uncertain multivariable systems obtained by system identification *International Journal of Systems Science* **45** pp 547-555
[4] Sanchez L, Alazard D, Michon G and Beriloz A 2014 Robustness Analysis of Helicopter Ground Resonance with Parametric Uncertainties in Blade Properties *Journal of Guiadance Control and Dynamics* **36** pp 272-281
[5] Astrom K and Hagglund T 1995 *Advanced PID control* (USA: ISA) **1** p 480
[6] Ángel L, Pérez M P, Díaz-Quintero C R and Mendoza C 2012 ADAMS/MATLAB co-simulation: Dynamic systems analysis and control tool *Applied Mechanics and Materials* **232**, pp 527-531
[7] Viola J and Angel L 2015 Identification, Control and robustness analysis of a robotic system using fractional control *IEEE Latin America Transactions* **13** (5) pp 1294-1302
[8] Doyle J Francis and B Tannenbaum A 1990 *Feedback Control Theory* (USA: Macmillan Publishing Co) **1** p 224

IOP Conf. Series: Materials Science and Engineering **138** (2016) 012009 doi:10.1088/1757-899X/138/1/012009

Modelling of uncertainness for a flow and level system

C Hernández[1], L Angel[1] and J Viola[1]

[1] Universidad Pontificia Bolivariana, Bucaramanga, Colombia

E mail: (carlos.hernandez, luis.angel, jairo.viola)@upb.edu.co

Abstract This paper presents the identification of uncertainness that affects the dynamics of a flow and level system. Initially, flow a level system is descripted. Then, family of plants is determined from the identification of dynamic model for different operating conditions. The uncertain model reflects the changes for different operating conditions when the output flow and storage tank dimensions are varied. Finally, the maximum multiplicative uncertain is calculated to define the desired controller specifications to achieve a robust stability and performance of the closed loop system.

1. Introduction

System dynamics is affected by external and internal factors that do not allow obtain an accurate mathematical model of the system, which difficult the control system design and the desired operation conditions [1]. These factors might be due to a system not modelled dynamics, uncertainness it the model parameters, external perturbations, system nonlinearities among others. Because of this, the uncertainness modeling of the system is important since it provides information about unknown dynamics against different operating conditions and ease the control system design to reach the robust stability and performance conditions [2]-[7]. Therefore, this paper proposes the development of a methodology for the parametric uncertainness identification for a flow and level system to obtain the closed loop specifications to guarantee the robust stability and performance of the system. This paper is structured as follows. First the flow and level system is descripted. Second system identification is performed to determinate the family of plants in the presence of parametric uncertainness and external perturbations of the system. Third, multiplicative uncertain for the family of plants is calculated. Fourth, maximum uncertain condition is modeled and the operating conditions to reach robust stability and performance are stablished. Finally, conclusions are presented.

2. System description

Flow and level system is composed by three tanks (TK1, TK2 and TK3), a servo-valve (FV1), two solenoid valves (FV2 and FV3), two manual valves (HV1 and HV2), a pipeline system and a pump (P). Also, system has an ABB control module for the implementation of basic control strategies. Upper tanks TK1 and TK2 let develop of level control strategies and TK3 tank works as deposit tank. Servo-valve (FV1) is the system actuator and works with a 4-20 mA signal. In this paper, control variable is the level in the tanks TK1 and TK2. Figure 1a shows the flow and level system and Figure 1b shows the P&ID diagram of the system.

(a)

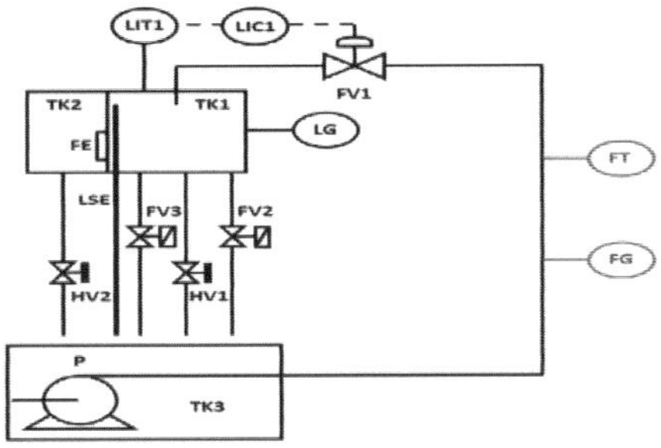

(b)

Figure 1. (a) Flow and level system and (b) P&ID diagram of the system

3. System identification

The identification of the mathematical model of the system is performed from the operation of the closed loop system. To do this, control loop was closed with a proportional controller with known gain. Data acquisition was performed with a NI DAQ 6008 and the model of the system was obtained using MATLAB identification toolbox, where input signal was the control action and the output signal belongs to the level in TK1.

4. Family of plants

Family of plants estimation was performed from the process identification for six operating conditions. These conditions using some combinations from the operation of solenoid valves FV2 and FV3 with TK1 and TK2 tanks. Table 1 shows the six operating conditions of the plant to be used for the identification process. ON condition in FV2 and FV3 indicates that solenoid valve is working, while ON condition in TK1 and TK2 indicates if the tank taken into account. Each condition in Table 1 is repeated five times for a total of 30 set of identification data.

Table 1. System operating conditions

Operating condition	FV3	FV2	TK2	TK1
1	ON	OFF	OFF	ON
2	OFF	ON	OFF	ON
3	ON	ON	OFF	ON
4	ON	OFF	ON	ON
5	OFF	ON	ON	ON
6	ON	ON	ON	ON

The family of plants for each operating conditions shown in Table 1 are identified with the transfer function model given by (1), which correspond to a second order with real poles system.

$$P(s) = k\frac{1 + T_z s}{(1 + T_{p1}s)(1 + T_{p2}s)} \tag{1}$$

Since for each operating condition data is taken five times. Table 2 shows an identification example of five models for the operating condition one.

Table 2. Family of plants for operating condition 1

K	Tp_1	TP_2	T_z	TK1
1,9829	55,079	25,248	8,1294	88,41
1,9955	74,975	11,028	2,8404	87,48
2,0005	87,785	4,3547	-2,2075	84,48
1,9754	76,355	11,111	2,312	83,69
2,0024	93,055	0,16539	-5,8687	81,84

Figure 2 shows the frequency response of the family of plants identified for the six operating conditions shown in Table 1. It shows that operating conditions variations affects the system dynamic at high frequencies. This variations in the system dynamics against operating conditions are modeled as parametric uncertainness of the model [8]. Nominal plant is calculated from the mean of parameters given by (1) for the 30 identified plants which is defined by (2) and is represented in Figure 2 by (*).

$$P_n(s) = 1.28\frac{1 + 0.25s}{(1 + 80s)(1 + 0.13s)} \tag{2}$$

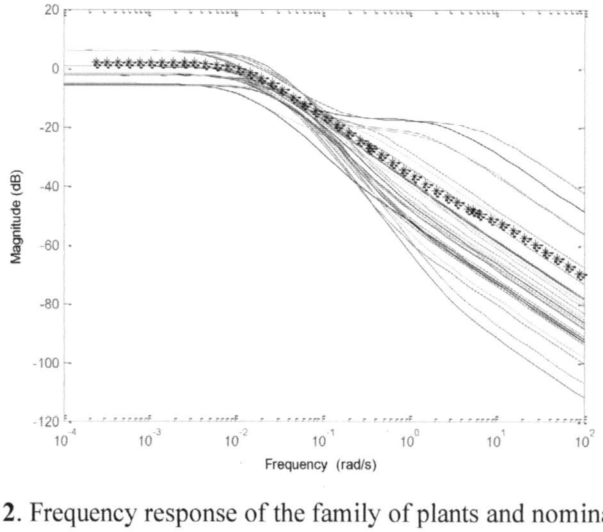

Figure 2. Frequency response of the family of plants and nominal plant.

5. Multiplicative uncertain modelling

From the family of plants, the multiplicative uncertain is modelled for each member of the family, which is defined by (3) and descripted in [9].

$$|\delta P_i(s)| = \left|\frac{\tilde{P}_i(s)}{P_n(s)} - 1\right| \tag{3}$$

where $P_n(s)$ correspond to the nominal plant descripted by (2) and $\widetilde{P_i}(s)$ to the i-th plant of the family of plants. Frequency response for the multiplicative uncertain of the family of plants is presented in Figure 3.

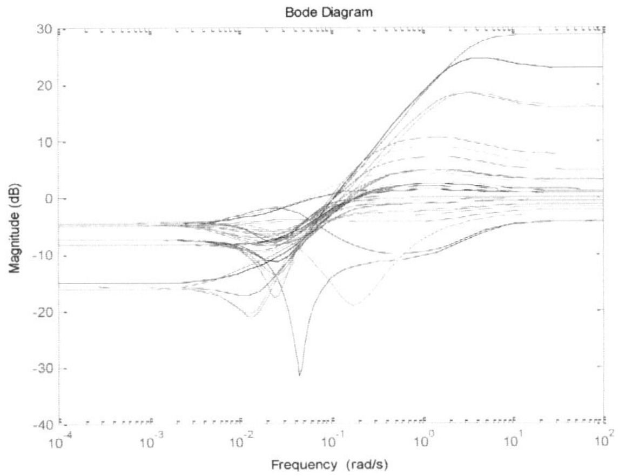

Figure 3. Multiplicative uncertain of the family of plants

Finally, the maximum multiplicative uncertain function is estimated according to the limit behaviors in low and high frequency giving the uncertain profile descripted by (4) and shown in Figure 4 represented by (*).

$$\delta_{pmax}(s) = 0.605\frac{(\frac{s}{0.0055}+1)(\frac{s}{0.105}+1)}{(\frac{s}{0.01}+1)(\frac{s}{3}+1)} \tag{4}$$

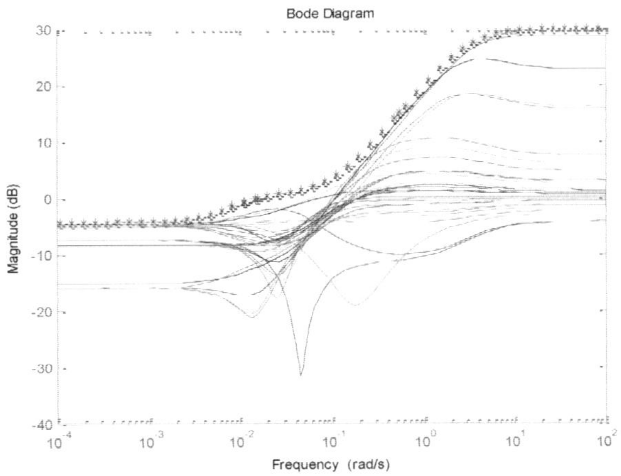

Figure 4 Maximum multiplicative uncertain profile

6. Robust operation conditions

From (4), closed operating conditions to achieve robust stability and performance are calculated according to (5).

$$|T(s)| \leq \left|\frac{1}{\delta P_{max}(s)}\right| \tag{5}$$

where $|T(s)|$ is the magnitude of the complementary sensibility function in closed loop and $\left|{}^{1}/_{\delta P_{max}(s)}\right|$ belongs to the magnitude inverse of the maximum uncertain profile of the family of plants given by (4). Closed loop system specifications are limited by the bandwidth $\beta\omega$ given by (5). Operating specifications comes from the desired characteristic polynomial defined by (6). The values for ζ and w_n are calculated from $\beta\omega \cong 0.64w_n$ for $\zeta = 0.7$ or $\beta\omega \cong w_n$ for $\zeta = 1$ [10]. As show in Figure 5, for the flow and level system the bandwidth correspond to $\beta\omega = 0.01$ rad/s, which guarantee that feedback system has a robust stability and performance [9].

$$s^2 + 2\zeta w_n s + w_n{}^2 \tag{6}$$

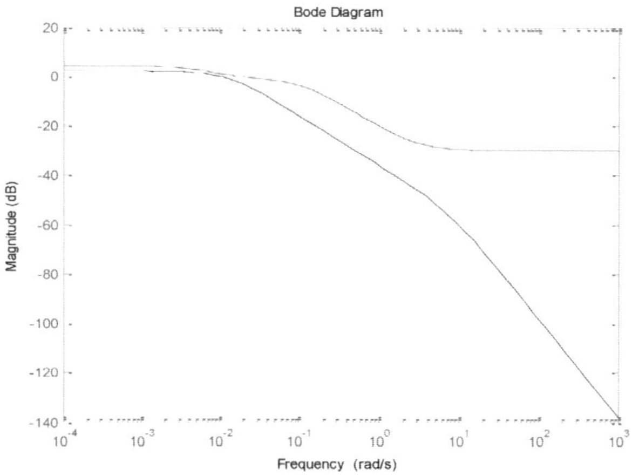

Figure 5 Bandwidth to achieve the conditions of robust stability and performance.

7. Conclusions

This paper presented a methodology for parametric uncertain experimental identification that affects the dynamics of a flow and level system. The family of plants was identified for different operating conditions of the system and the maximum multiplicative uncertain was calculated. Frequency response of the inverse maximum multiplicative uncertain allows to find the maximum system bandwidth to define the closed loop operation conditions of the system that met the condition of robust stability and performance. The obtained results show the dynamic behavior changes of the system when external perturbation appears. Thus, applying the proposed methodology is possible to modeling this changes through a multiplicative uncertain function to define the desired robust performance conditions in the controller design stage.

References

[1] Astrom K and Hagglund T 1995 Advanced PID control (Sweden:Pearson)
[2] Khadraoui S, Nounou H, Nouno M, Datta A and Bhattacharyya S 2015 A nonparametric approach to design robust controllers for uncertain systems: Application to an air flow heating system *Journal of Process Control* **36** pp 1-10
[3] Tufa L and Ramasamy M 2011 Closed-loop identification of systems with uncertain time delays using ARX–OBF structure *Journal of Process Control* **21**(8) pp 1148-1154
[4] Mahanta C and Senthilkumar D 2010 Identification of uncertain nonlinear systems for robust fuzzy control *ISA Transactions* **49** (1) pp 27-38
[5] Alfaro-Ponce M, Argüelles A and Chairez I 2014 Continuous neural identifier for uncertain nonlinear systems with time delays in the input signal *Neural Networks* **60** pp 53-66

[6] Hasiewicz Z, Mzyk G and Śliwiński P 2012 Mixed parametric-nonparametric identification of Hammerstein and Wiener systems - a survey IFAC Proceedings Volumes **45** (16) pp 464-469
[7] Salloum R, Arvan M and Moaveni B 2013 Identification, uncertainty modelling and robust controller design for an electromechanical actuator *International Conference on Control, Automation, Robotics and Embedded System (CARE)*
[8] Rodríguez Rubio F and López Sánchez M J 1996 Control Adaptativo y Robusto (España: Universidad de Sevilla)
[9] Doyle J and Tannenbaum A 1991 Feedback Control Theory (US: Macmillan Publishing)
[10] Landau I D and Zito G 2006 Digital Control System (Germany: Springer)

IOP Conf. Series: Materials Science and Engineering **138** (2016) 012010 doi:10.1088/1757-899X/138/1/012010

Smart Grids and their Applicability for the Development of the Electricity Sector for Colombia in the year 2050

J Viola[1] and C Aceros[1]

[1]Professor, Universidad Pontificia Bolivariana, Bucaramanga, Colombia
Email: (jairo.viola, cesar.aceros)@upb.edu.co

Abstract. Smart Grids are a technology that can be used to implement a sustainable energy scheme of a country. Therefore, this paper proposes the development of a prospective analysis of Smart Grids as a tool to ensure energetic security in Colombia in 2050. Using LEAP software, a base scenario for Colombian energy demand has developed according to current policies, with a time horizon from 2012 to 2050. The energy analysis is based on three scenarios, taking into account the impact of cogeneration in the residential and industrial sector using renewable energy and the power quality indicators. The results show that the implementation of Smart Grids generate energy savings and increasing the coverage of the national electricity system, ensuring energetic security of the country by 2050.

1. Introduction

Smart grid are the next step in the development of a country electrical grid, because it allows the implementation of electricity cogeneration systems and the inclusion of renewable energies contributing to the diversification of energetic matrix [1], [2], [3]. In the case of Colombia generates 70% of electricity by hydroelectricity. This shows a big dependence of hydric resources, which will be affected for a change in the precipitation levels due to the global warming [4]. Above situation will carry an energy default, which must be satisfied with new clean energy sources that contribute to reduce the greenhouse gases emissions and ensure the energetic security of the Colombian Citizens. Therefore, in this paper is performed a prospective analysis using a scenario-based methodology for the smart grid in Colombia to the year 2050, which contribute to reach the energetic security and the diversification of the Colombian energy matrix [5], [6]. The paper is structured as follows. First, the parameters for smart grid evaluation are defined, which are power quality, distributed cogeneration and energetic security. Second, the Colombian energetic model is presented. Third, the three scenarios, Trend (ET), climate change (CC) and technological alternative (TA) are presented and implemented in the prospective software LEAP. Fourth, the behaviour of electric demand and electricity generation are analysed for each scenario, evaluating the compliance with the parameters mentioned above. Finally, conclusions are shown.

2. Smart Grid

Smart grid presents a new paradigm for the development of generation, distribution, marketing and consumption tasks [1], [2], [3], integrating the information and communication technologies (ICT) to the power management system, allowing the implementation of control schemes and energy management flowing through the grid. Figure 1 and Figure 2 shows the difference between a conventional power distribution network and an intelligent network.

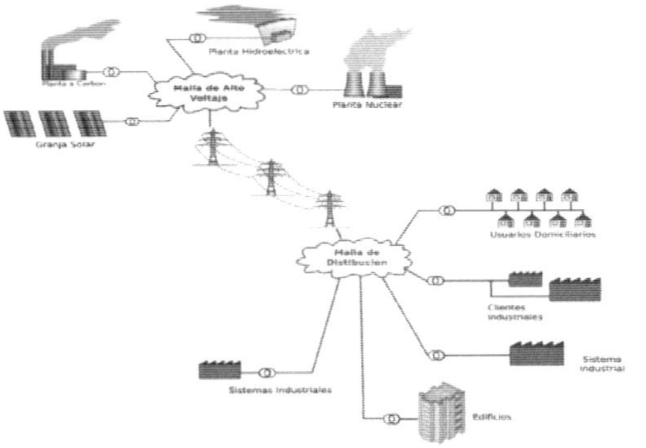

Figure 1 Conventional power grid

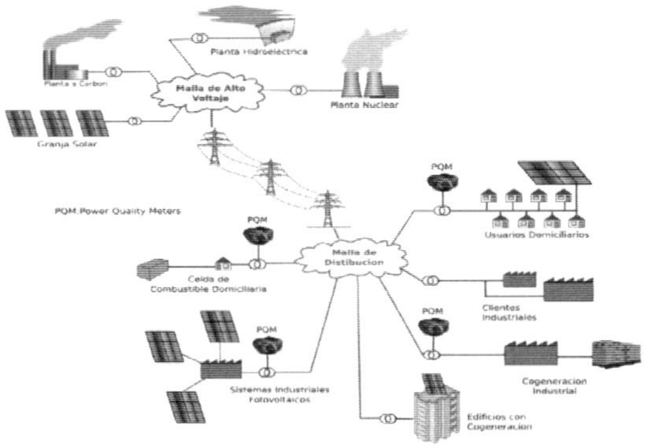

Figure 2 Smart grid

As shown in Figure 1, the conventional electricity grid conforms by big generators that produce energy, which is transmitted via a distribution system from the generator to final consumers who are the residential and industrial sectors. This indicates that the energy flows in one direction. Moreover, Figure 2 shows the outline of a smart grid, in which big generators do not exclusively produce energy, but each user on the network can produce their own energy using renewable energies as solar panels and wind turbines, injecting power to the grid. Energy regulation is made by a control centre responsible for distributing energy efficiently according to the required demand. Identification of the robotic-joint dynamic model.

3. ASSESSMENT PARAMETERS FOR THE SMART GRID EVALUATION IN THE ENERGY PROSPECT

For the evaluation of the feasibility, impact and contribution of smart grids in the ET, TA and CC scenarios, the following parameters were employed.

3.1. Power Quality

Power Quality defines as all the inherent parameters of electricity waveform such as amplitude, frequency, wavelength, continuity; which to assess deviations in the voltage waveform and the effects of this on the equipment connected to the grid.

3.2. Distributed Generation

Distributed generation is one of the main features of smart grids, which presents the possibility for the power network users, both domestic and industrial to generate and inject energy to the grid using alternative power schemes as solar panels or wind turbines, large or small scale. This article will evaluate cogeneration, taking into account the levels of production of electricity from solar and wind renewable energy in each scenario.

3.3. Energy security

The energy security is a responsibility of governments, whose purpose is to ensure the power supply to the inhabitants of a country, by protecting energy infrastructure, diversification of sources of electricity supply, tending to ensure fair and regulated prices, always looking for energy sources with the least environmental impact. To evaluate the energy security, the types of generation technology employees in each scenarios to meet the energy demand of the country will be analysed which must comply with the above features.

4. COLOMBIAN ENERGY MODEL

Energy planning software LEAP is used for modelling the Colombian energy system [7], [8], which is based on the model proposed by the mining and energy-planning unit (UPME). This model is the technical-economic. That is, considered the most important macroeconomic and social variables with the technical aspects of Colombia's energy matrix, such as energy reserves, the demand for each of them in different sectors, the processes of transformation of primary energy such as oil or water in fuel and electricity respectively, considering the losses and efficiency of these processes.

4.1. Colombian energy model features
The main characteristics of technical-economic model of the energy system for Colombian electricity sector (demand generation, transformation and loss) are:

4.2. Macroeconomic and demographic variables
As macroeconomic variables, the Colombia's gross domestic product (GDP) and the benefit of each of the sectors of the Colombian industry are considered. As demographic variable population growth rate, which according to the national statistics unit (DANE) is 1.3% to 2050. Each of these variables has as starting point for the development of prospective analysis referred to the time series since 2008 until 2012, projected until 2050.

4.3. Electricity demand
Electricity demand takes into account the residential, industrial, transport, commercial, mining, construction, agriculture and consumption sectors. According to the information provided by the UPME, it will have a total demand of electricity historic sectors for 2012 of 44.44 TWh.

4.4. Generation and transformation processes
At this stage, model considers the necessary processes for the transformation of primary energy such as coal, wood, oil water or electricity to meet demand for electricity. Energy conversion processes, the percentage share in power generation and energy losses of the transformation process are allow for each energetic.

5. Scenario approach
5.1. Trend scenario
This scenario takes as its starting point the baseline scenario raised by the UPME in [8], assuming a Constant annual population growth of 1.5% until 2050, an Increase in GDP of 3.4% annually, a Primary energetic for electricity production for 2050 will be the water resource and Industrial growth will continue according to forecasts of industry and trade ministry.

5.2. Climate Change Scenario
In this scenario, climate change will alter the hydrological conditions of Colombia, decreasing precipitation levels throughout the country, generating a deficit of electric power approximately 10,000 Gwh in 2050. The characteristics of this scenario are a Continuous annual population growth of 1.5% until 2050, with a GDP increased by 3.4% annually. Since water resources will be affected by climate change, the shortage of electricity is supplemented by energy sources based on coal and natural gas. The level of efficiency of thermal processes does not increase despite new technologies and global trends. Finally, Alternative energies have a minimum stake in the Colombian energy matrix.

5.3. Technological alternative scenario
In this scenario the same considerations submitted to the scenario of climate change will be made, however, in the alternative renewable energy technology (solar and wind) arise in place of fossil fuels as sources of energy to meet the shortfall in demand Colombia electricity in 2050.

6. OBTAINED RESULTS
Using LEAP, the conditions for each scenario are calculated. The proposed scenarios (TE, CC, TA) are compared with the UPME scenarios World electric and energy efficiency scenarios posed in [8]. In each

of the scenarios, parameters show in section 3 were assessed by analysing the production of electricity in Colombia 2050.

6.1. Generation of electricity in the scenarios posed

6.1.1 Trend Scenario

The generation of electricity in the baseline scenario shown in Figure 3, which shows the power generation by 2050 will be 152 Twh, which, 86% will produced using hydric resources. Meanwhile the share of renewable energies will be at 2.3% and fossil fuels with a share of 11%.

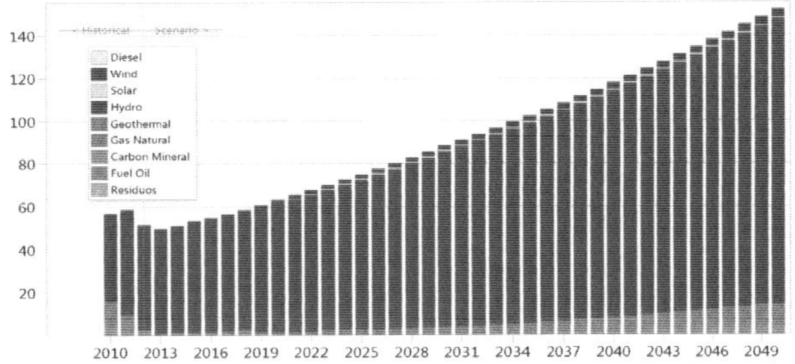

Figure 3 Electricity generation trend scenario in Twh

6.1.2 Technological alternative scenario

The electric power generation technology in alternative scenario shown in Figure 4. In this figure, the power generation by 2050 will be 150 Twh, which shows that 38% will be produced using water resources. Meanwhile the share of renewable energies will be in a 59.55% and fossil fuels with a lower share of 3%.

6.1.3 Climate Change Scenario

The generation of electricity in the Climate Change scenario shown in Figure 5. Which shows the power generation by 2050 will be 150 Twh, which shows that 41% will be produced using fossil fuels. Meanwhile the share of renewable energies will be at 0.26% and hydroelectricity will have a share of 56%.

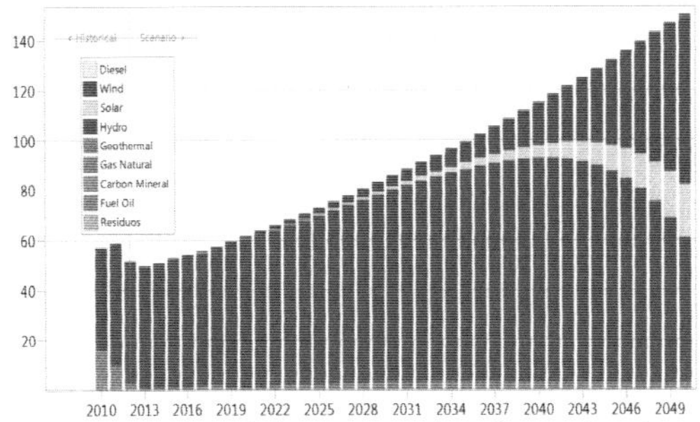

Figure 4 Electricity generation for technology alternative scenario in Twh

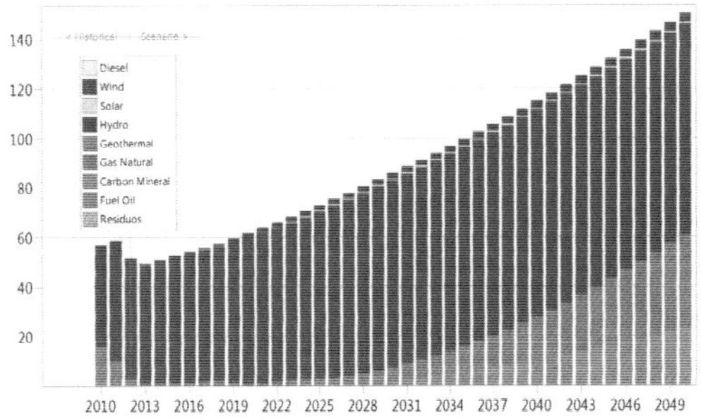

Figure 5 Electricity generation for climate change scenario in Twh

6.2. Generation of electricity in the scenarios proposed by the UPME

The scenarios presented below correspond to those raised by the UPME in [5], which electric world and energy efficiency are having into account as the most important for the analysis of smart grids.

6.2.1 Electric World Stage

The generation of electricity in the baseline scenario shown in Figure 6, the electric power generation by 2050 will be 241 Twh, which will be produced in a 78.63% using water resources. Meanwhile the share of renewable energies will be at 1.86% and fossil fuels with a share of 19.51%.

6.2.2Energy efficiency scenario

The generation of electricity in the baseline scenario shown in Figure 7. As can be seen, the power generation by the year 2050 will be 41.4 Twh where 86% will be produced using water resources. Meanwhile the share of renewable energies will be in a 13.11% and fossil fuels with a lower share of 1%.

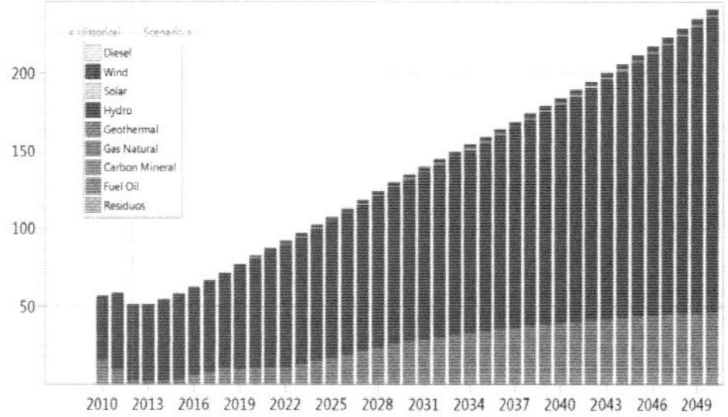

Figure 6: Electricity generation electrical world stage in Twh

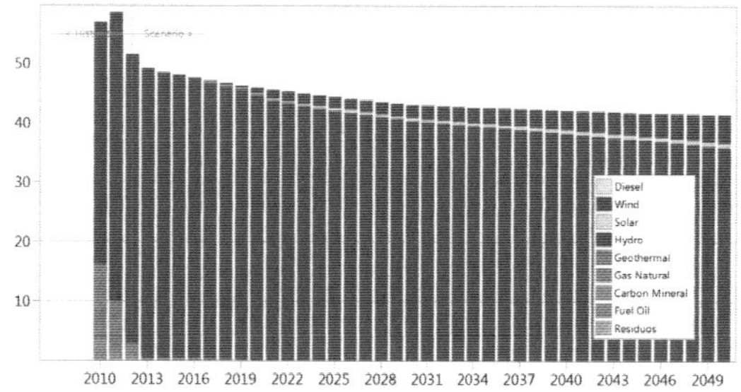

Figure 7: Electricity generation for energy efficiency scenario in Twh

7. Result analysis

From the results obtained in Section 6, we can see that in the energy prospects of Colombia 2050, electricity production in all scenarios is covered mainly by hydro, leaving second alternative energies (solar and wind). However, the threat of climate change for Colombia is quite high, since according to [8], a change in rainfall patterns will cause a shortage of water resources throughout the country. In Figure 8 the deficit of electricity to be presented by department for 2050 is presented.

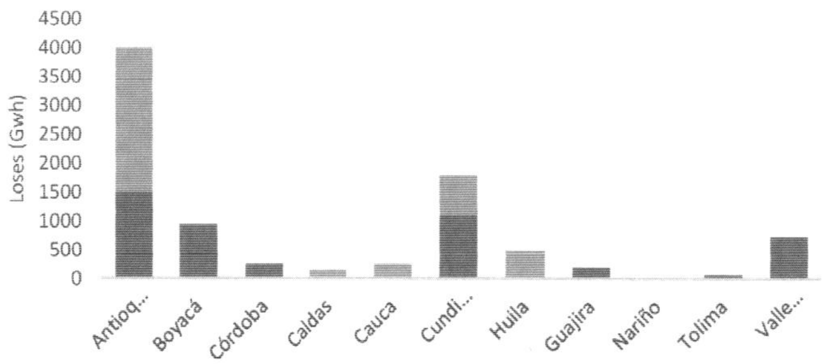

Figure 8: Deficit electricity department in Colombia by 2050 due to climate change (green High risk and blue medium risk)

According to Figure 8, if the level of climate change is high or medium, the department of Antioquia will present a deficit of up to 2,500 GWh due to the scarcity of water resources. Considering the above for smart grids evaluation according to the parameters described in Section 3. In the electric world or energy efficiency scenarios, it will not be possible to guarantee a constant and uninterrupted service of electricity to system users. this implies a lower power quality, thus making energy security of citizens is not ensured, since not having a large share of renewable energies, implementing rule cogeneration schemes urban, rural and industrial, making it difficult the implementation of smart grids in these scenarios. In the climate change scenario, if a part of this deficit is supplied using fossil fuels, again not resort to cogeneration schemes to replace lost due to water scarcity, but instead used energy based on fossil fuels , which have a greater environmental impact, in breach of the principle of further reducing emissions from energy security. Thus in this scenario not intelligent network scheme implemented is observed. In the case of alternative technological scenario, the implementation of renewable energy, cleanly contribute to reducing emissions of greenhouse gases and mainly supply the energy deficit due to climate change. Importantly, these alternative sources can be installed in residential, rural and industrial sectors, making possible the small-scale generation in a distributed, decentralized system of power generation. Therefore, it can be said at this stage the implementation of smart grid for constant monitoring of system power and a fulfilment centre that controls the distribution of energy according to the demand for electricity brought at any time of the day will be needed. Further reducing dependence on water sources, will allow using this resource in other activities such as human consumption or agricultural production, optimizing its use.

8. Conclusions

This paper describes the development of a prospective analysis for the evaluation of smart grids and its application in Colombia 2050 employing a scenario based methodology, using as evaluation parameters the energy security, power quality and distributed generation. For this, the Colombia energy model is modelled using the energy planning software LEAP. The proposed scenarios in LEAP model are compared with the UPME scenarios. From the obtained results, the main power source in Colombia is the hydroelectricity, which in a Climate Change (CC) scenario will affect the energy security and power quality in Colombia in 2050 due to the hydric resource default. Thus, the smart grid turns into an option to avoid the energy deficit by integrating renewable energies in the Colombian energy matrix. Also, the scenario based prospective model is a tool that helps to predict the energy trends for a country. In the Case of Colombia, this model allows to understand the behaviour of the electric sector to create policies and regulations that be robust against climate, technological or social changes in Colombia until 2050, ensuring the energy security and power quality.

9. ACKNOWLEDGEMENT

Special thanks to the mining-energy planning unit UPME of Colombia as facilitators of information necessary for the implementation of the energy model of Colombia for this study.

References

[1] Singhal Anjali and Saxena R P 2012 Software models for Smart Grid *International Workshop on Software Engineering for the Smart Grid (SE4SG)* pp 42-45
[2] Ramakrishnan K 2015 Smart grid implementation in India — A case study of Puducherry Pilot Project, India *International Conference on Energy Economics and Environment (ICEEE)* pp 2-27
[3] Sinha A et al Smart grid initiative for power distribution utility in India 2011 *IEEE Power and Energy Society General Meeting* pp 1-8
[4] Gallego J D, Franco C J and Zapata S 2015 Policies for the utilization of hydropower potential in Colombia using small plants *IEEE Latin America Transactions* 13(12) pp 3844-3850
[5] Tapia A, Bernal C, Rodriguez G and Soto J D 2013 Energy efficiency and environmental impact of small markets in Barranquilla-Colombia *12th International Conference on Environment and Electrical Engineering (EEEIC)*
[6] Gaona E E, Trujillo C L and Guacaneme J A 2015 Rural microgrids and its potential application in Colombia Renewable and Sustainable Energy Reviews **51** pp 125-137
[7] Siteur J 2014 *the long-range energy alternatives planning model (leap) and wood energy planning overview and exercises* (Stockholm environmental institute)
[8] Antonio M and Camargo C 2015 *plan energético nacional Colombia : ideario energético 2050* (unidad de planeación minero energética)
[9] Unidad de planeación minero energética 2015 *plan de expansión de referencia generación – transmisión* pp 475
[10] Bonilla U M and Héctor H H 2013 *vulnerabilidad y opciones de adaptación del sector eléctrico colombiano frente al cambio climático* (Unidad de planeación minero energética)

Modeling of lighting behaviour of a hybrid lighting system in inner spaces of Building of Electrical Engineering

L Amado[1], G Osma[1] and R Villamizar[1]

[1] Department of Electrical, Electronics and Telecommunications Engineering, Universidad Industrial de Santander, Carrera 27 calle 9, Bucaramanga, Colombia

E-mail: german.osma@correo.uis.edu.co

Abstract. This paper presents the modelling of lighting behaviour of a hybrid lighting system - HLS in inner spaces for tropical climate. HLS aims to mitigate the problem of high electricity consumption used by artificial lighting in buildings. These systems integrate intelligently the daylight and artificial light through control strategies. However, selection of these strategies usually depends on expertise of designer and of available budget. In order to improve the selection process of the control strategies, this paper analyses the Electrical Engineering Building (EEB) case, initially modelling of lighting behaviour is established for the HLS of a classroom and an office. This allows estimating the illuminance level of the mixed lighting in the space, and energy consumption by artificial light according to different lighting control techniques, a control strategy based on occupancy and a combination of them. The model considers the concept of Daylight Factor (DF) for the estimating of daylight illuminance on the work plane for tropical climatic conditions. The validation of the model was carried out by comparing the measured and model-estimated indoor illuminances.

1. Introduction

The artificial lighting in buildings is a significantly energy load [1]. The HLS can reduce this consumption, since these are based on control techniques taking advantage of the daylight reducing the use of artificial lighting [2]. Studies have demonstrated the benefits of the HLS; [3] shows an energy saving of 30% in Torino (Italy) by using fluorescent lamps and [4] presents a HLS used in Shanghai (China), which is analyzed for fluorescent and LED lamps, where energy consumption is reduced approximately 41% and 66%, respectively. Because of the potentiality of the HLS in the buildings, it is necessary continue with the conducting studies to intensify the obtaining of benefits. Such is the case of modeling of lighting behavior in inner spaces for the design, control and optimization of HLS.

In that sense, this paper addresses the modeling of lighting behavior and energy consumption of two study cases of current HLS in Electrical Engineering Building (+ 7.1 ° latitude) of the Industrial University of Santander (Bucaramanga, Colombia).

This work was conducted in seven stages. First stage is a brief review of the literature (Section 2); second stage consists of the fundamentals of modeling obtained (Section 3); third step is the description of the study cases (Section 4); fourth stage addresses the measurement process in the selected spaces (Section 5); fifth and sixth stages present the formulation of models of lighting behavior and energy consumption, respectively (Sections 6 y 7); finally, seventh stage shows the results of the validation of the models of the lighting behaviour.

2. Literature review

The lighting behavior can be described from models representing the conditions of a space and the relationship with several parameters and variables. For example, Daylight Factor (DF) is a relationship between the interior daylight and external daylight. This factor has been modeled of different ways, including a matrix as a function of the geometric position in the space [5], [6]. Likewise, Parise *et al.* [5] model the interior daylight as a function of the geometric position and DF in the room. Moreover, the total interior illuminance is modeled as the combining of the artificial light and daylight on the working plane [6], [7]. Similarly, Kamaruzzaman *et al.* [8] evaluated the performance of the lighting in an office building in the tropical zone from of the modeling of artificial lighting, daylighting, position of the sun, and kind of sky, among others factors.

Other factors related to indoor lighting are subject of modeling, as illuminance on the work surface, control, and energy consumption. Parise *et al.* [5] describe the control lighting strategies by a set of equations. Fischer *et al.* [9] present the relationship between the dimming levels of luminaries and the illuminance on the working plane. Moreover, Roisin *et al.* [10] show a linear relationship between the power consumption by lamp and its luminous flux. While, Osma *et al.* [11] and Ozenc *et al.* [12] determine polynomial functions relating the power consumption to the dimming level and the energy consumed according to dimming levels. Depending on the purpose of the modeling, it can be mathematical or analytic type, based on computational tools as Matlab [6]. Also, it can be developed from statistical methods (linear regression [10], multivariate non-linear regression [13]), artificial intelligence techniques (machine learning [9], neural network [14], fuzzy logic [7]), and through lighting simulation tools (Radiance [9], Dialux, [10], Virtual Enviroment (VE) [8], Daysim [10]).

3. Fundamentals of modeling

The total illuminance (ET) (lux) on an interior point of the working plane for a room is the sum of all contributions of interior daylight (E_D) (lux) and artificial illuminance (E_A) (lux), being the last one provides by luminaires, which may be or not managed by a lighting control strategy [6]. Therefore, the total illuminance on an interior point can be defined as:

$$ET(t,x,y) = E_A(t,x,y) + E_D(t,x,y) \tag{1}$$

Where the artificial illuminance $E_A(t,x,y)$ is a function of distance existing between horizontal projection of mid-point of luminaire until mesh point (measurement point of the illuminance). ITs value depends on daylight availability. Such adjustment is performed by a lighting control strategy. $E_D(t,x,y)$ is a function of geometric position and varies as a function of the time [5]. It is calculated by equation (2) for each point (x,y) of the space. Where $E_E(t)$ (lux) is the external daylight illuminance and $DF(\%)_{(i,j)av}$ is the daylight factor average for each point (x,y) of room.

$$E_{D(i,j)}(t) = E_E(t) \cdot DF(\%)_{(i,j)av} \tag{2}$$

The $DF(\%)_{(i,j)av}$ estimation considers these aspects: i) In tropical places as Bucaramanga, DF must to be obtained with respect to incident daylight at the façade and not at the roof, since vertical external daylight can provide more accurate information than the horizontal case. Furthermore, in a lit side room, daylight may be more nearly proportional to the amount of daylight falling on the window, rather than to the external horizontal daylight [8]; ii) DF must be calculated by source, so must not be combined side sources (windows of different facades) nor zenithal sources (solar tubes).

4. Description of the study cases

This paper presents two study cases of HLS applied in the Electrical Engineering Building (EEB) of Industrial University of Santander. The first study case is a classroom located in the third floor (see Figure 1), which has a HLS consisting of an lighting control by on/off switching of lamps in response to occupancy and daylight availability (one occupancy sensor including an on/off photocell OSC10-

M0W), twelve fluorescent luminaires (4x14W T5 – electronic ballast), and windows at the south façade. The second case is a meeting room located in the fifth floor (see figure 2); its HLS is composed of a lighting control by dimming (10 % - 100%) of the lamps in response to occupancy and daylight availability (one occupancy sensor OSC10-M0W and dimming photocell LS-301), four fluorescent luminaires (4x14W T5 – dimming ballast), and windows at the south façade.

Figure 1. Class room (11,9 m x 6,6 m).

Figure 2. Office (meeting room) (5,4 m x 3,6 m).

5. Measurement process

The matrix models are based on measurements of daylighting and artificial lighting, which were taken with light meters according to a mesh for a height of 0.75m (working plane). The measurement mesh consists of a point set established in accordance to the dimensions of space.

The figure 3 and figure 4 show the mesh of measurement of selected spaces, with 18 points (3x6) and 9 points (3x3), respectively. The artificial illuminance was measured at night, whilst during daytime were taken the measurements of interior daylight illuminance and exterior daylight illuminance (E_{E1}, E_{E2}) for calculating to DF. Each point of mesh is an array element ($e_{i,j}$).

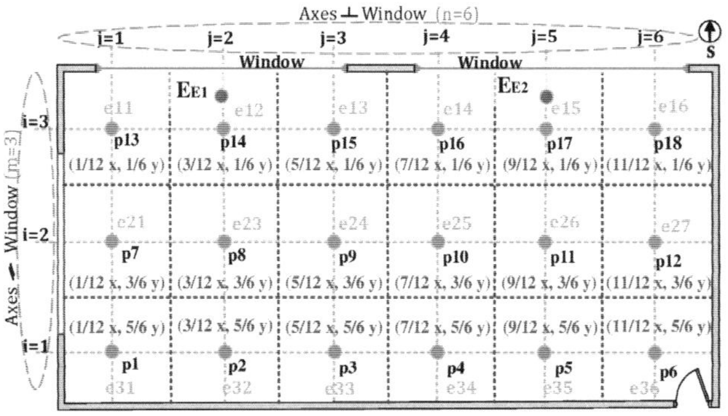

Figure 3. Measurement mesh of classroom.

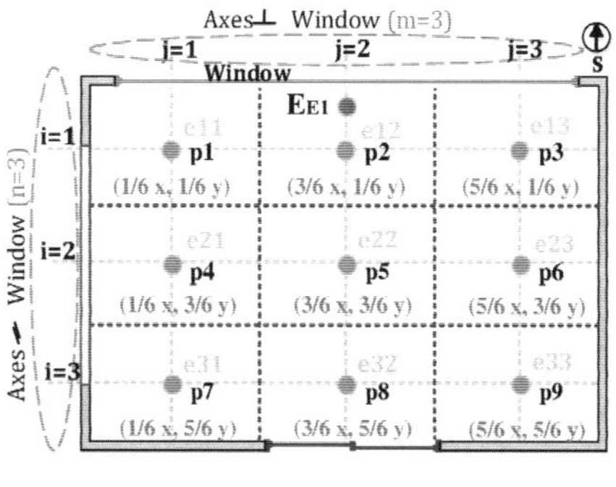

Figure 4. Measurement mesh of office.

6. Model of lighting behavior

6.1. Estimation of Daylight Factor (DF)

DF is the percentage ratio of interior daylight illuminance to external daylight illuminance [8]. In this paper, DF was calculated as an average value in each point $DF(\%)_{(i,j)av}$. The matrix model of DF for two spaces is shown by the equation (3), where $E_{D(i,j)}(t)$and $E_{E(i,j)}(t)$ are the interior daylight illuminance (in the absence of artificial light) and the external daylight illuminance, respectively; which are measured for every hour of sunshine in each measurement points and n is the number of hours considered for calculating of $DF(\%)_{(i,j)av}$.

$$DF(\%)_{(i,j)av} = \left(\frac{\sum_{t=1}^{n} \frac{E_{D(i,j)}(t)}{E_{E(i,j)}(t)}}{n} \right) \cdot 100 \qquad (3)$$

Results indicate that DF tends to be the same during daytime and varies according to geometric position inside the room, as shown in figure 5 and figure 6. The highest values of DF are presented close to the window and decreases as they move away from this.

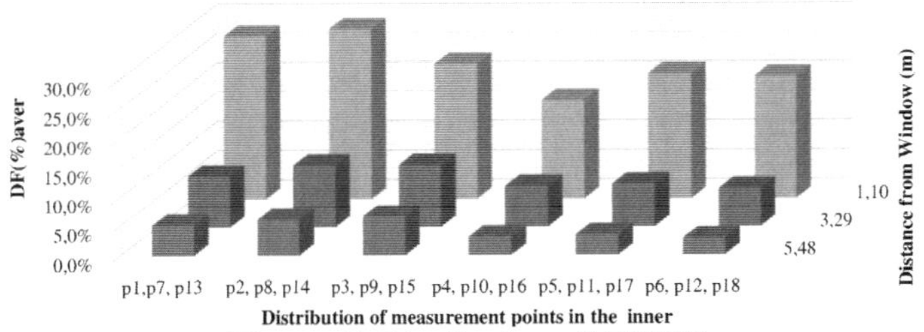

Figure 5. Distribution of $DF(\%)$ in the class room

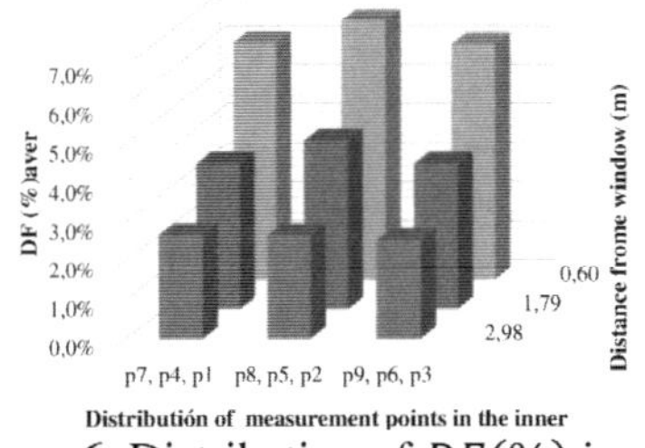

Figure 6. Distribution of $DF(\%)$ in the meeting room

6.2. Daylight illuminance

The interior daylight illuminance $E_{D(i,j)}(t)$ (lux) in each point is estimated for every time instant by means of the matrix model of illuminance given by equation (4).

$$E_{D(i,j)}(t) = E_{E(i,j)}(t) \cdot DF(\%)_{(i,j)av} \tag{4}$$

6.3. Artificial illuminance

The interior artificial illuminance is measured in lux for each point of classroom $E_{A\_CR(i,j)}$ (lux) and of the meeting room $E_{A\_MR(i,j)}$ (lux). The artificial illuminance matrices are presented below.

$$E_{A\_CR(i,j)} = \begin{bmatrix} 422 & 440 & 448 & 468 & 461 & 422 \\ 500 & 542 & 536 & 528 & 533 & 482 \\ 659 & 697 & 680 & 702 & 717 & 618 \end{bmatrix} \qquad E_{A\_MR(i,j)} = \begin{bmatrix} 322 & 461 & 301 \\ 588 & 821 & 535 \\ 553 & 763 & 530 \end{bmatrix}$$

6.4. Total interior illuminance - $ET_{(i,j)}(t)$

The equation (5) allows the calculation of $ET_{(i,j)}(t)$ (lux) for every time t and every point of the mesh; where $E_{A\_CS(i,j)}(t)$ is the artificial illuminance defined by the lighting control strategy of each space.

$$ET_{(i,j)}(t) = E_{D(i,j)}(t) + E_{A\_CS(i,j)}(t) \tag{5}$$

6.5. Artificial illuminance according to control strategy

The values of $E_{A\_CS(i,j)}(t)$ depends on adjustment done by lighting control strategy. This adjustment is a function of factors such as: state of switching of luminaires, availability of daylight, occupancy, minimal value of daylight and strategy of lighting control. $E_{A\_CS(i,j)}(t)$ for the classroom varies between zero and the values of artificial lighting matrix according to following conditions:

$$E_{A\_CS(i,j)}(t) = 0 \begin{cases} Oc_t = 0; \\ Oc_t = 1 \wedge Oc_{t-1} = 0 \wedge e_{nat\_pc}(t) > e_{min}; \\ Oc_t = 1 \wedge Oc_{t-1} = 1 \wedge lum_{t-1} = \text{"Off"} \wedge e_{nat\_pc}(t) > e_{min} \end{cases} \tag{6}$$

$$E_{A\_CS(i,j)}(t) = E_{A(i,j)} \begin{cases} Oc_t = 1 \wedge Oc_{t-1} = 1 \wedge lum_{t-1} = \text{"On"}; \\ Oc_t = 1 \wedge Oc_{t-1} = 1 \wedge lum_{t-1} = \text{"Off"} \wedge e_{nat\_pc}(t) < e_{min} \end{cases} \tag{7}$$

Where Oc_t and Oc_{t-1} are the current state and previous state of the occupancy, respectively; $e_{nat\_pc}(t)$ is the minimal value of daylight illuminance in the space; e_{min}= 400 lux is the minimal value of illuminance that must be ensured according to the RETILAP [15]; lum_{t-1} is the On/Off state of luminaires for $t-1$. The artificial illuminance of the meeting room $E_{A\_CS(i,j)}(t)$ is defined by the required artificial illuminance $E_{A\_req(i,j)}(t)$ (lux) to achieve a value of e_{min} equal to 435 lux. $E_{A\_req(i,j)}(t)$ is calculated from of the following conditions:

$$E_{A\_req(i,j)}(t) = e_{min} - MIN(e_{D21}: e_{D23}) \{Oc = 1 \wedge MIN(e_{D21}: e_{D23}) < e_{min} \quad (8)$$

$$E_{A\_req(i,j)}(t) = 0 \begin{cases} Oc = 0; \\ Oc = 1 \wedge MIN(e_{D21}: e_{D23}) > e_{min} \end{cases} \quad (9)$$

$$E_{A\_req(i,j)}(t) = MIN(\mathrm{E}_{\mathrm{A}_{\mathrm{SJ(i,j)}}}) \{Oc = 1 \wedge MIN(e_{D21}: e_{D23}) = 0 \quad (10)$$

Where $MIN(e_{D21}: e_{D23})$ is the minimum value of daylight that can be given on the working plane (table of meeting). The second component is the relationship between artificial illuminance according to control in percentage and the dimming voltage (0 - 10 V), which is represented by the equation (11).

$$E_{A\_CS} = -0{,}0024DV^3 + 0{,}031DV^2 + 0{,}022DV + 0{,}0606. \quad (11)$$

Considering $E_{A\_req(i,j)}(t)$ and the equation (11), it can be determined the dimming voltage DV, from which the matrix $E_{A\_CS(i,j)}(t)$ of meeting room can be estimated by equations (12) and (13).

$$E_{A\_cs(i,j)}(t) = 0 \begin{cases} E_{A\_req(i,j)}(t) = 0; \\ E_{A\_req(i,j)}(t) = \frac{e_{A\_req23}}{e_{A23}} \leq 0{,}01 \end{cases} \quad (12)$$

$$E_{A\_cs(i,j)}(t) = (-0{,}0024DV^3 + 0{,}031DV^2 + 0{,}022DV + 0{,}0606) \cdot E_{A(i,j)} \left\{ \frac{e_{A\_req23}}{e_{A23}} > 0{,}01 \right. \quad (13)$$

7. Modeling of energy consumption

7.1. Class room

The energy consumption of lighting system is determined by the equation (14). Where P_{lu_t} is the power of the luminaries (80W), n_{lu} is the number of luminaires (12), h_T is the total hours of operation of the luminaires, P_{OS} is the occupancy sensor power (1 W) and n_{OS} is the number of occupancy sensors (2). In general, the operating daily time is 10 hours, although the artificial lighting is on only 6 hours because the control strategy uses the daylighting; thus the daily energy consumption is near to 5,8 kWh.

$$E_{cT} = P_{lu\_t} \cdot n_{lu} \cdot h_T + P_{OS} \cdot n_{OS} \cdot 24hr \quad (14)$$

7.2. Meeting room

The meeting room has 4 dimming luminaries of 78W. The power consumption of each luminary (P_{lu}) varies according to DV and is given by the equation (15). The total energy consumption of lighting system of the meeting room is given by the equation (16). Where $EC_{t_{lu}}$ is the total energy consumed by the luminaires, EC_{t_OS} is the total energy consumed by occupancy sensor, and EC_{t_fdim} is the total energy consumed by dimming photocell.

$$P_{lu} = -0.108DV^3 + 1.217DV^2 + 3.274DV + 31{,}403 \quad (15)$$

$$E_{c\_total} = EC_{t_{lu}} + EC_{t\_OS} + EC_{t\_fdim} \quad (16)$$

The daily values of EC_{t_OS} and EC_{t_fdim} are always 24Wh, while the $EC_{t_{lu}}$ varies for every hour according to values of DV and availability daylighting. Considering a typical daily conditions work, as is described by Osma et al. [11] (previous work about these spaces), it is possible to estimate that energy consumption is around 0,6 kWh/day.

8. Validation

In order to validate the matrix model, a comparison between the measured and the estimated total interior illuminances $\boldsymbol{ET}_{(i,j)}(\boldsymbol{t})$ was done for of the meeting room at 3:22 p.m. (to see Figure 7). Furthermore, was calculated the normalized mean squared error (NRMSE) and mean absolute error (MAE), whose values was 6,9% and 15,96 lux, respectively. Based on the presented comparison between the measured and the estimated total interior illuminances and the error measures (MAE and RMSE), the validation of the model can be considered as successful.

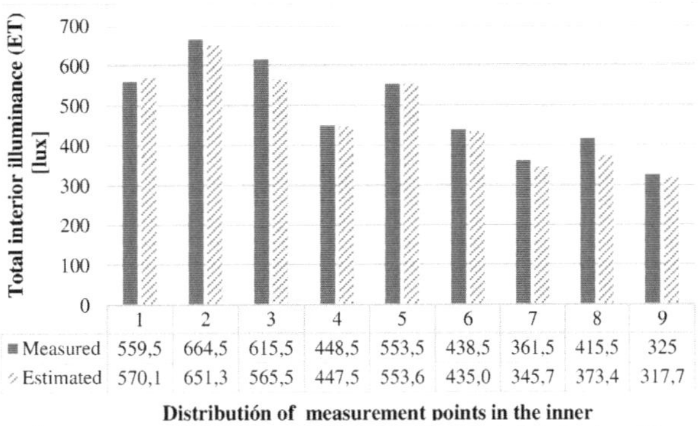

	1	2	3	4	5	6	7	8	9
■ Measured	559,5	664,5	615,5	448,5	553,5	438,5	361,5	415,5	325
Estimated	570,1	651,3	565,5	447,5	553,6	435,0	345,7	373,4	317,7

Figure 7 Comparing the measured and model-estimated total interior illuminance at 3:22 p.m.

9. Conclusions

The study has shown that the presented model of total interior illuminance $\boldsymbol{ET}_{(i,j)}(\boldsymbol{t})$ can be used as a satisfactory method for estimating the illuminance in inner spaces.

The proposed models of lighting behavior and energy consumption are simple mathematical equations that can be useful for the study of HLS. For example, in the development of tools allowing to select of optimal way the components (sensors, photocells) of control strategy of the HLS, according to criteria such as minimal financial cost and minimal energy consumption, among others.

References

[1] Haq M A U, Hassan M Y, Abdullah H, Rahman H A, Abdullah M P, Hussin F and Said D M 2014 *Renew. Sustain. Energy Rev.* **33** 268-279

[2] Tetri E and Bhusal P 2010 Annex 45 *Guidebook on Energy Efficient Electric* (Finlandia: Aalto University) pp 132-134

[3] Aghemo C, Blaso L and Pellegrino A *Autom. Constr.***43** 10-22

[4] Ge A, Qiu P, Cai J, Wang W and Wang J 2014 *Appl. Opt.* **53** 1869–73

[5] Parise G, Martirano L and Member S 2013 *Ind. Appl. IEEE Trans.* **49** 242–249

[6] Chen X, Xiao H, Fang J and Peng L 2015 *The 27th Chinese Control and Decision Conf.* Qingdao (IEEE) pp 804–809

[7] Logar V, Kristl Ž and Škrjanc I 2014 *Energy Build.* **70** 343-351

[8] Kamaruzzaman S N, Edwards R, Zawawi E M A and Che-Ani A I 2015 *Energy Build* **90** 85–93

[9] Fischer M, Wu K and Agathoklis P 2012 *32nd Int. Conf. on Distributed Computing Systems Workshops* Macau (IEEE) pp 245-249

[10] Roisin B, Bodart M, Deneyer A and D'Herdt P 2008 *Energy Build.* **40** 514-523

[11] Osma G A, Ordóñez G, Amado L Y and Villamizar R 2014 *Appl. Mech. Mater.***704** 368–372

[12] Ozenc S, Uzunoglu M and Guler O 2014 *Energy Build.* **77** 432–439

[13] Lo Verso V R M, Pellegrino A and Pellerey F 2014 *Energy Build.* **76** 151–163

[14] Wang Z and Tan Y K 2013 *Energy Build.* **62** 514–521

[15] Ministerio de Minas y Energía 2010 *RETILAP Anexo General Reglamento Técnico de Iluminación y Alumbrado Público* (Bogota: Ministerior de Minas y Energía) chapter 4

Characterization of hybrid lighting systems of the Electrical Engineering Building in the Industrial University of Santander

D Galvis[1], C Exposito[1], G Osma[1], L Amado[1] and G Ordóñez[1]
[1]Electrical Engineering Department, Industrial University of Santander, Bucaramanga, Cra 27 calle 9, CO

E-mail: german.osma@correo.uis.edu.co

Abstract. This paper presents an analysis of hybrid lighting systems of Electrical Engineering Building in the Industrial University of Santander, which is a pilot of green building for warm-tropical conditions. Analysis of lighting performance of inner spaces is based on lighting curves obtained from characterization of daylighting systems of these spaces. A computation tool was made in Excel-Visual Basic to simulate the behaviour of artificial lighting system considering artificial control system, user behaviour and solar condition. Also, this tool allows to estimate the electrical energy consumption of the lighting system for a day, a month and a year.

1. Introduction

Hybrid lighting systems – HLS combine daylight and artificial light. Its operation is focused on control of artificial lighting in response to natural illuminance for a given time [1], [2]. Currently, it is evident that there is a significant range of devices for implementing such systems. However, the number of studies about the detailed operation is reduced, particularly about daylighting scenarios.

Therefore, it is necessary to carry out several analysis of the daylighting behavior in inner spaces. This can be performed by means of Daylight Factor - DF, which is one percentage indicator calculated as the ratio between the illuminance at a point inside and the outside illuminance.

This paper presents the lighting characterization for some inner spaces of the Electrical Engineering Building (EEB) by calculating DF and artificial lighting measurements. EEB is a green building pilot of the Industrial University of Santander (Bucaramanga, Colombia), and works as a laboratory for studying systems aimed at reducing energy consumption in buildings for warm tropical climate.

The lighting curves for each space is based on the artificial lighting level for nights, the set point of operation of the photocells and the contribution of natural light (DF). Additionally, a study was conducted for selecting the outer reference for DF calculation, outdoor terrace (horizontal) and front (vertical) was performed. To display the information collected a computational tool in Excel (ALE3T) was made, with which can be estimated the lighting performance and energy consumption according to space, time, date, level of solar radiation incident.

This study is presented in this document as follows: Section II describes general considerations, Section III exposes the methodology and results, and finally, Section IV presents the Conclusions.

2. General considerations

2.1. Hybrid lighting systems
Daylight is really potent by itself, and it is possible to improve it using smart architecture as an aesthetic method and technical tool, guaranteeing excellent luminance levels and comfort [3]. But, even so, it can get better using hybrid lighting system concept. HLS are groups of techniques and equipment that respond to the natural lighting contribution by artificial lighting. These are configured to respond to the contribution of natural light. Because of this, design should consider that incident solar radiation on windows depends on the type of sky; additionally, the daylighting distribution is as a function of technical characteristic of the inner space. However, this study is not done in detail. Figure 1 shows a typical HLS.

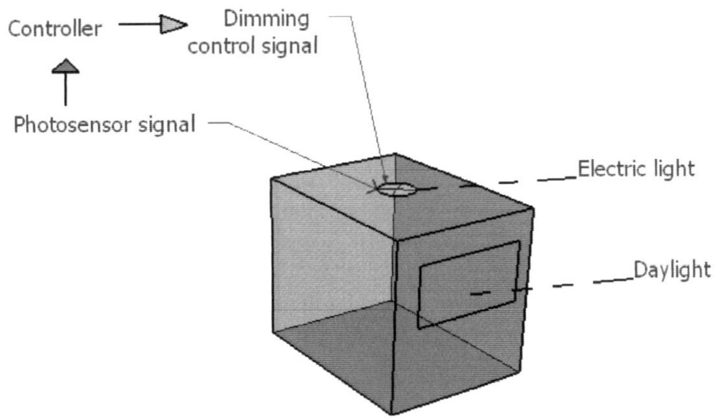

Figure 1. Typical hybrid lighting system.

These systems provide benefits such as cost savings in billing for lower energy consumption, and reduction of the carbon footprint in constructions. For example, Chow *et al.* [4] tested a daylight harvesting system with T5 fluorescent lamps for individual offices adjoining the atrium in Hong Kong, achieving an energy saving from 59% to 75%. These good practices improve the level of sustainability in buildings while maintaining comfort and facility of use [5].

In Colombia, the technical regulation about lighting is called "*Reglamento Técnico de Iluminación y Alumbrado Público*" – RETILAP. Its Section 450 demands that new nonresidential buildings (>500m^2) must have lighting control systems, specifically use of natural lighting and control artificial lighting [6].

2.2. Daylight factor
The integration of hybrid lighting systems is based on natural lighting. The measurable parameters regarding lighting can be dimensional (illuminance), dimensionless factor (uniformity) and percentage (Df and glare) [6]. Daylight Factor (Df) is the term applied to the percentage of the division between internal illuminance produced by natural light up to the working plane (Eint) and the illuminance on the outside (Eext), determined in the same instant with a uniformly overcast sky without obstructions [6]. In this particular case (classroom) talking about Colombia, *Df* value must be minimum 2% [6]. *Df* calculus can be so easy as equation (1) or can considering a lot of parameters like the relative air mass or the luminous turbidity as show in [7], [8] . For this paper was taken standard equation:

$$Df = \frac{E_{int}}{E_{ext}} * 100\% \quad (1)$$

E_{ext} is typically attributed to the horizontal illuminance measured in terrace, such as shown in [7], [9]; however, it is not reliable for tropical zone due to the sun movement. Specifically, it is possible to evidence that if the horizontal solar radiation is the same for two different dates in the year, the Df will be different. This is because the solar height can be significantly different, and therefore varies the amount of solar radiation that may enter through the openings. Therefore, the Df was considered with reference to the facade.

3. Methodology

This research is conducted in three stages: (i) the illuminance measurement of natural and artificial lighting in selected areas, (ii) the construction of the virtual model of the EEB in DIALux, and (iii) the creation of the virtual tool to estimate illuminance and energy consumption.

The study was performed in Electrical Engineering Building from the Industrial University of Santander, which is located in the city of Bucaramanga (Lat. 7.1°N and Long. 71.3°W). This building is 969 meters above sea level in a tropical zone. It has an average hours of sunlight 12.1 and solar radiation of 4.8 $kWh/m^2/day$ [10]. Integrated hybrid lighting strategies in EEB are presented in Table 1.

Table 1. Current hybrid lighting strategies in EEB.

Strategy	Luminaries	Control
Type I	Fluorescent On/Off	Occupancy sensor with ON/OFF photocell
Type II	Fluorescent On/Off	Occupancy sensor and ON/OFF photocell
Type III	Dimming fluorescent	Occupancy sensor and dimming photocell
Smart hybrid I	Dimming fluorescent	Occupancy sensor and dimming photocell, controlled by BAS (Building Automation System)
Smart hybrid II	Dimming fluorescent and solar tubes	Occupancy sensor, dimming photocell and solar tubes, controlled by BAS (Building Automation System)

3.1. Light metering

The measurement step consisted of meshing of spaces, daylighting measurement, data collection, Df estimation, artificial illuminance measurement and estimation of the set point in spaces. The measurement was performed with lux meters. Meshing is the location of points according to the sizes of the spaces. Figure 2 depicts an example of meshing for a space.

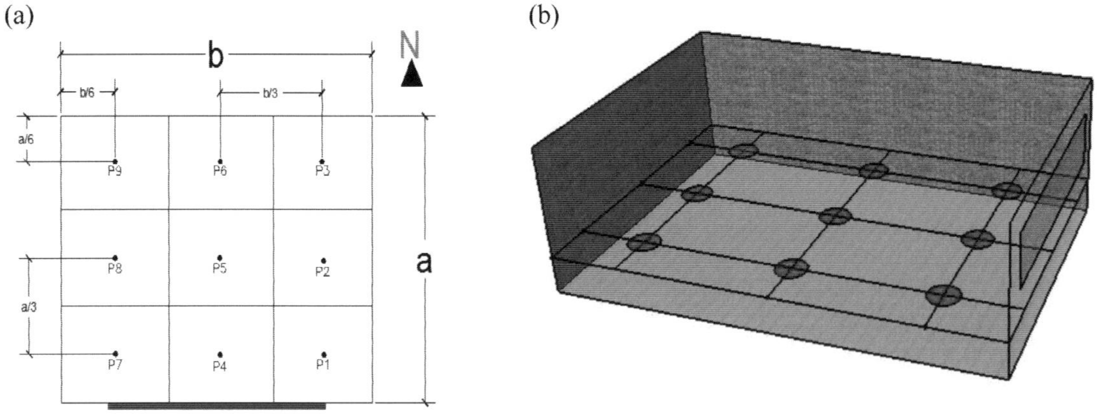

Figure 2. Example of used meshing. (a) 2D meshing; (b) 3D meshing.

Daylight is monitored from 08:00 until 18:00 according to the mesh for a height of 75 cm; simultaneously, the incident illuminance on façade was measured. The incident illuminance on the terrace was obtained by measuring the horizontal global solar radiation obtained from the weather station and using a conversion factor (115 lux/W/m^2). Later, the Df was calculated for the two references, terrace and facade.

Figure 3 shows the results obtained for the classroom 206. Figure 3. (a) ilustrates the high variability Df when is calculated according to the illuminance on the terrace. Conversely, Figure 3. (b) shows a tendency uniform for Df calculated according to the illuminance on the façade, being the best option. Figure 4 shows a 3D representation of Df for the classroom 206 mesh.

Measurement of artificial lighting was performed during evening time (after sunset) (> 6:00 pm) considering the established meshing. Next, the operating set point of the photocells is estimated, which is the minimum value of daylighting contribution that can exist without the need to turn on the lamps.

3.2. Virtual building development

The virtual model of the building was made with the DIALux software (v4.12). The data from simulations describes similar results to analysis based on measurements. The variables analyzed were artificial illuminance and Df. The virtual model allows to study several scenarios in order to predict lighting behavior. These information is considered by ALE3T to estimate the energy consumption of the lighting system.

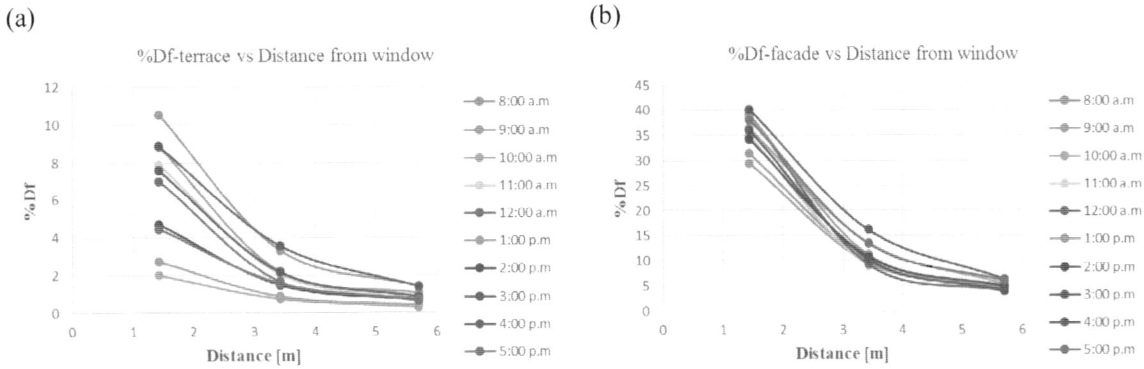

Figure 3. Hourly Df in classroom 206 according to distance to window for two references: (a) Df regard to terrace; (b) Df regard to facade.

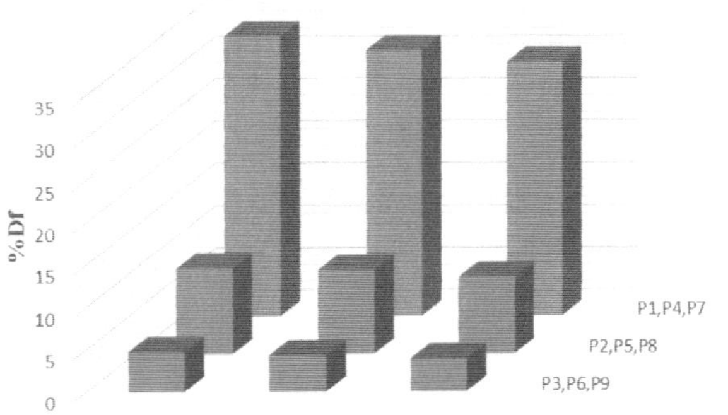

Figure 4. Performance of Df-facade in classroom 206 (8:00 am).

3.3. Computational tool ALE3T

There are some techniques or programs to determinate Daylight, such as shown in [11], [12]. For this tool, Daylight is determined by *DF*. It means, it takes a reference DF (Measured), and this is multiplied by horizontal illuminance measured in terrace.

ALE3T is a computational tool developed in Microsoft Excel through Visual Basic Editor. This aims to estimate the power consumption by artificial lighting in specific spaces of Electrical Engineering Building. Also, it can display demand curves, hybrid illuminance and Df. The Figure 5.(a) presents the main menu of the tool, where is possible to define the variables, such as space, occupancy, set point for photocells, month and horizontal solar radiation. While Figure 5.(b) is a snapshot of the graphical results.

The first graph ("Average illuminance by source for each hour of the day") shows the artificial illuminance (green line), which is activated when there is occupancy (black line) and average natural illuminance (red dots) is below set point level (orange line). Also, the minimum natural lighting in the space (celestial points) and combined illuminance (purple line) is shown. The second graph allows to know the energy performance, through the power curve (red line) and energy consumption (blue line).

(a) (b)

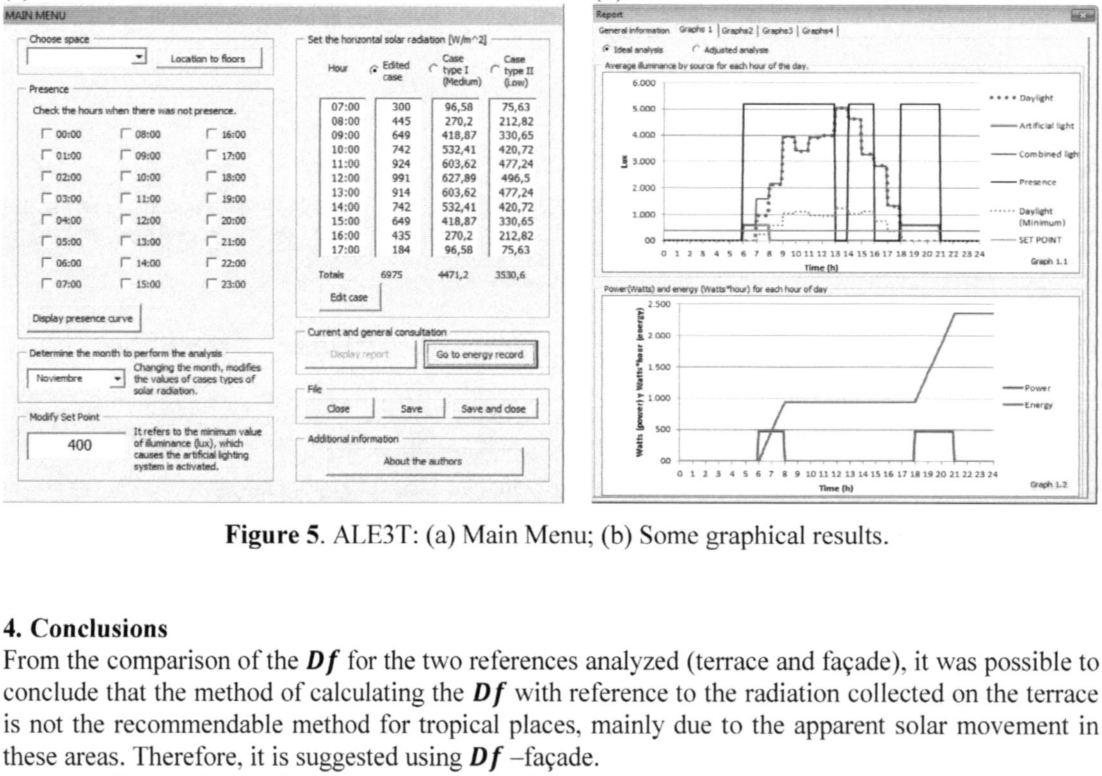

Figure 5. ALE3T: (a) Main Menu; (b) Some graphical results.

4. Conclusions

From the comparison of the $\boldsymbol{Df}$ for the two references analyzed (terrace and façade), it was possible to conclude that the method of calculating the $\boldsymbol{Df}$ with reference to the radiation collected on the terrace is not the recommendable method for tropical places, mainly due to the apparent solar movement in these areas. Therefore, it is suggested using $\boldsymbol{Df}$ –façade.

According to the performance curves of artificial lighting, it was possible to deduce that the illuminance of space is not uniform and varies depending on the configuration the luminaire system and the degree of reflection from the walls and windows.

The computational tool ALE3T fulfills its main objective that is to estimate the energy consumption of the lighting system, and exposes the lighting curves for inner spaces of EEB. The tool should be improved considering aspect such as: time analysis of 5 minutes (currently the analysis is hourly), presentation scenario where the curtain is used and the lamps are partially turned off, and including of the sun tubes (zenithal light).

Comparing the five hybrid lighting strategies in EEB, it found the major power savings between them. This corresponds to labeled in this article Smart hybrid II, because in the same scenario this one consumes less energy, obtaining saves near to 50% respect to worse strategy (Type I), because the consumption are around 2,4 kWh/day and 4,8 kWh/day, respectively.

References

[1] Parise G and Martirano L 2013 Combined Electric Light and Daylight Systems Ecodesign *IEEE Trans. Ind. Appl* **49** 1062-1070

[2] Yin W, Xiao H, Liu S, Yan X and Kang Q 2014 Intelligent Lighting Control System Using Illumination Sensor for Office Building *Sensor Letters* **12** 236-240

[3] Hansen E, Hammershoj Olesen G and Mullins M 2013 Home Smart Home: A Danish Energy-Positive Home Designed With Daylight *Proceedings of the IEEE* **101** 2436-2449

[4] Chow S, Li D, Lee E and Lam J 2013 Analysis and prediction of daylighting and energy performance in atrium spaces using daylight-linked lighting controls *Applied Energy* **112** 1016-1024

[5] Dubois M and Blomsterberg Å 2011 Energy saving potential and strategies for electric lighting in future North European, low energy office buildings: A literature review *Energy and Buildings* **43** 2572-2582

[6] Ministerio de Minas y Energía. República de Colombia 2010 *Reglamento Técnico de Iluminación y Alumbrado Público RETILAP* (Bogota, Colombia)

[7] Li D and Cheung G 2006 Average daylight factor for the 15 CIE standard skies *Lighting Research and Technology* **38** 137-152

[8] Li D, Lam T and Wu T 2012 Estimation of average daylight factor under obstructed CIE Standard General Skies *Lighting Research and Technology* **46** 187-197

[9] Mangkuto R 2015 Validation of DIALux 4.12 and DIALux evo 4.1 against the Analytical Test Cases of CIE 171:2006 *LEUKOS* 1-12

[10] Eosweb.larc.nasa.gov 2015 NASA Surface meteorology and Solar Energy: RETScreen Data Online: https://eosweb.larc.nasa.gov/sse/RETScreen/

[11] Ng E, Gadi A, Wong F, Mu J, Gadi A and Lee M 2007 Predicting daylight availability dynamically based on forecasts of a weather observatory *Lighting Research and Technology* **39** 69-80

[12] Geebelen B, van der Voorden M and Neuckermans H 2005 Fast and accurate simulation of long-term daylight availability using the radiosity method *Lighting Research and Technology* **37** 295-312

IOP Conf. Series: Materials Science and Engineering **138** (2016) 012013 doi:10.1088/1757-899X/138/1/012013

Building energy analysis of Electrical Engineering Building from DesignBuilder tool: calibration and simulations

J Cárdenas[1], G Osma[1], C Caicedo[1], A Torres[1], S Sánchez[1], and G Ordóñez[1]
Department of Electrical, Electronic Engineering and Telecommunications, Industrial University of Santander, Cra. 27 Calle 9, Bucaramanga, Colombia.

E-mail: german.osma@correo.uis.edu.co

Abstract. This research shows the energy analysis of the Electrical Engineering Building, located on campus of the Industrial University of Santander in Bucaramanga – Colombia. This building is a green pilot for analysing energy saving strategies such as solar pipes, green roof, daylighting, and automation, among others. Energy analysis was performed by means of DesignBuilder software from virtual model of the building. Several variables were analysed such as air temperature, relative humidity, air velocity, daylighting, and energy consumption. According to two criteria, thermal load and energy consumption, critical areas were defined. The calibration and validation process of the virtual model was done obtaining error below 5% in comparison with measured values. The simulations show that the average indoor temperature in the critical areas of the building was 27°C, whilst relative humidity reached values near to 70% per year. The most critical discomfort conditions were found in the area of the greatest concentration of people, which has an average annual temperature of 30°C. Solar pipes can increase 33% daylight levels into the areas located on the upper floors of the building. In the case of the green roofs, the simulated results show that these reduces of nearly 31% of the internal heat gains through the roof, as well as a decrease in energy consumption related to air conditioning of 5% for some areas on the fourth and fifth floor. The estimated energy consumption of the building was 69 283 kWh per year.

1. Introduction

The residential and commercial buildings have energy consumption between 20% and 40% of total world energy consumption [1]. This has become a problem due to increased energy demand and related emissions. In order to reduce consumption in buildings, initially it is required to understand the energy performance of the building. This task is complex and therefore demands the use of energy simulation tools [2].

In the tropics energy performance of a building is highly influenced by the outside weather conditions [3]. High values of temperature, relative humidity, and solar radiation are characteristic of cities like Bucaramanga. Such conditions can lead to heating of the interior spaces, which results in significant increases in cooling loads and reduce the sensation of thermal comfort of the occupants[3]–[8].

Currently, the Industrial University of Santander studies the energy performance of the Electrical Engineering Building for warm tropical weather conditions. This building incorporates energy-saving strategies such as solar tubes, green roof, natural ventilation, natural lighting, and highly efficient hybrid systems (lighting and air conditioning). Such infrastructure allows conducting studies aimed at identifying factors, which influence significantly on the energy performance and user comfort, especially in areas exposed to critical conditions of environment, electrical charges and occupation.

Considering the above, the objective of this research is to carry out an energy analysis of the Electrical Engineering Building through simulations obtained through DesignBuilder.

This document consists of three sections, Section II describes the process of obtaining and validating the virtual model of the building, Section III presents the main results of the energy simulations, and Section IV exposes the conclusions about energy analysis.

2. Methodology

The methodology for this research is based on process modeling and energy analysis of buildings in energy simulation tools as DesignBuilder [9]. Initially, a description of the building under study is performed in order to present the most relevant features of the building have been taken into account in the energy modeling. Subsequently, the process of virtual modeling and calibration through environmental measurements is shown. Finally, a compilation of energy simulations carried out for calibration and energy analysis of the virtual model is shown.

2.1. *Building under study*

Electrical Engineering Building - EEB is located on the central campus of the Industrial University of Santander in Bucaramanga - Colombia, a region of tropical climate. The building has an area of 2700 m^2 distributed in five floors. The building includes spaces for classrooms, study rooms and offices. The ambient temperature in Bucaramanga varies between 24°C and 30°C and annual rainfall levels are close to 1279 mm. Figure 1 shows the building.

Figure 1. South façade of the Electrical Engineering Building.

The envelope of the building is composed of outer walls in concrete, brick and drywall. All areas of significant occupation of the building they have glazing areas over 40% in the walls of the facade, allowing the use of natural ventilation and lighting. Passive strategies for sustainable construction like green roof and solar tubes were implemented in the flat roof of the building. Hybrid systems (lighting and air conditioning) were defined in some areas of the building.

2.2. *Virtual model of building*

The virtual model of the building in energy simulation tool DesignBuilder consisted of geometric modelling and incorporating data model. The geometric model was carried out in four stages: generation blocks through DXF plans; tracing interior partitions to create zones; definition of windows, doors and other openings; and modelling of architectural details and buildings surrounding area. Modelling solar tubes required a special approach for including in model. The whole geometric model of EEB is shown in Figure 2.

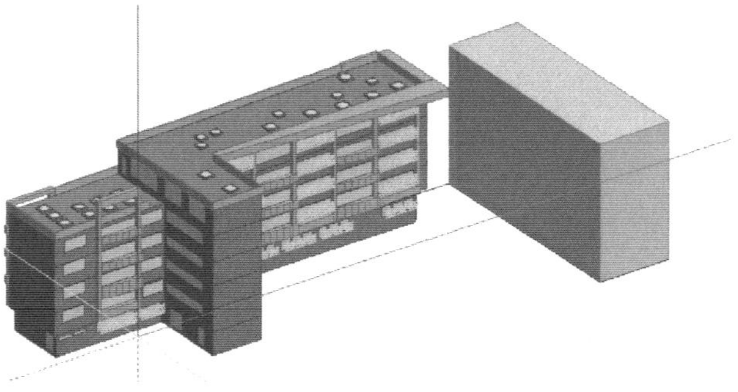

Figure 2. Geometric model of Electrical Engineering Building created in DesignBuilder.

Filling model data in DesignBuilder is performed considering five data categories: activity, constructions, openings, lighting and HVAC systems. Activity data are related to the use of zones and conditions of users comfort. The materials and thicknesses of the various components of the envelope were defined in the construction data category. The characteristics of the windows and doors were taken into account in the openings data category, of the same way that lighting and HVAC information.

2.3. *Model calibration*

The calibration process involves the identification of critical zones, environmental variables monitoring, and the validation of the model.

2.3.1. Identification of critical zones. The identification of critical zones was carried out taking into account criteria how: occupation, artificial lighting loads, operating loads of equipment, ventilation and infiltration loads, and solar radiation loads. A quantitative assessment of the impact on energy consumption and thermal comfort have the criteria mentioned above was carried out for all building spaces. Aspects such as the number of people, weekly time intensity and metabolic activity were considered for the evaluation of the occupation criteria. The installed power lighting and power equipment in each area were taken into account in the evaluation of lighting and equipment loads. For ventilation and infiltration loads were considered the presence in the spaces of windows, sun breakers, vents and heat extractors. Finally, the location of the areas was taken into account in assessing the loads associated with solar radiation. Table 1 shows the areas identified as critical zones.

Table 1. Critical zones of Electrical Engineering Building.

Floor	Zone
1	Group study room and Single study room
4	Classroom 402 Classroom 404, and 406 Classroom.
5	Student waiting room, Meeting room, Waiting room, File area and project and service zone.

2.3.2. Environmental variables monitoring. The evaluation of the accuracy of the results of the simulations are based on the measurement of environmental variables such as air temperature, relative humidity, air velocity, and daylighting level in each critical zone. The measurements were performed between 8:00 am and 6:00 pm from Monday to Saturday during June of 2014. Table 2 shows the average values measured for each of the variables mentioned above.

Table 2. Mean values measured air temperature, relative humidity, air velocity and level of daylighting in each of the zones identified as critical.

Zone	Air temp. [° C]	Relative humidity [%]	Air velocity [m/s]	Average level of daylighting [lux]
Project and servicie zone	25.7	69.6	0.2	2213.2
Student waiting room	27.1	67.3	0.1	1742.2
File area	26.2	67.5	0.1	1216.1
Meeting room	25.8	68.8	0.3	797.1
Waiting room	26	69.4	0.3	1673.8
Classroom 402	26.5	69.1	0.2	1910.8
Classroom 404	26	64.9	0.6	2907.6
Classroom 406	26.8	69.6	0.2	1106.6
Group study room	26.7	68.8	0.3	959.9
Single study room	26.9	70.1	0.1	586.2

The measured value of total annual energy consumption of the building was also considered. This corresponds to 48 200 kWh per year.

2.3.3. Validation of the model. The validation process consisted of a comparison between measured and simulated values of air temperature, relative humidity, air velocity, and daylighting level and energy consumption.

Differences minor that 5% were found in the most of the critical zones. This value suggests that the virtual model of the Electrical Engineering Building is acceptable for use in energy analysis, the main objective of this research. The greatest differences were found in the area of higher occupancy (Group study room) for the air temperature and air velocity with 12.9% and 7.14%, respectively. In the case of relative humidity, the difference found was close to 0.2%.

The simulated annual energy consumption was close to 50,000 kWh, this value is approximately 4% higher than the measured value. A tuning process carried out with respect to occupation and use of systems and equipment, in order to mitigate the percentage differences.

2.4. *Simulation plan*

Two sets of simulations mainly comprise the analysis of the energy performance of the building: Preliminary simulations and energy simulations. Preliminary simulations were done for calibrating the virtual model, from which air temperature, relative humidity, air velocity, and daylighting level were obtained for comparison with the measured values. The energy simulations were used to estimate the energy consumption of the building and the heat gains.

3. Analysis of results

In order to describe the energy performance of the building, the results of the energy simulations related to thermal comfort, indoor heat gains and the impact of energy saving applications were analyzed.

3.1. *Thermal comfort*

Average annual values of temperature and relative humidity were taken into account in establishing the internal thermal conditions in the main building critical areas. According to them, only the critical zones with air conditioning have thermal comfort conditions all year. The most unfavourable environmental conditions were found for Group study room and Classroom 402, where the air temperature has average annual values of 30.31 °C and 29.96 °C. In the case of relative humidity, an average annual value of 67% was found.

The thermal conditions of the building are characterized in DesignBuilder using thermal comfort indices. According to these, the most representative critical zones without air conditioning have a thermal indoor environment among slightly hot and very hot (values between 1 and 3 on the scale of comfort indices) as shown in Figure 3.

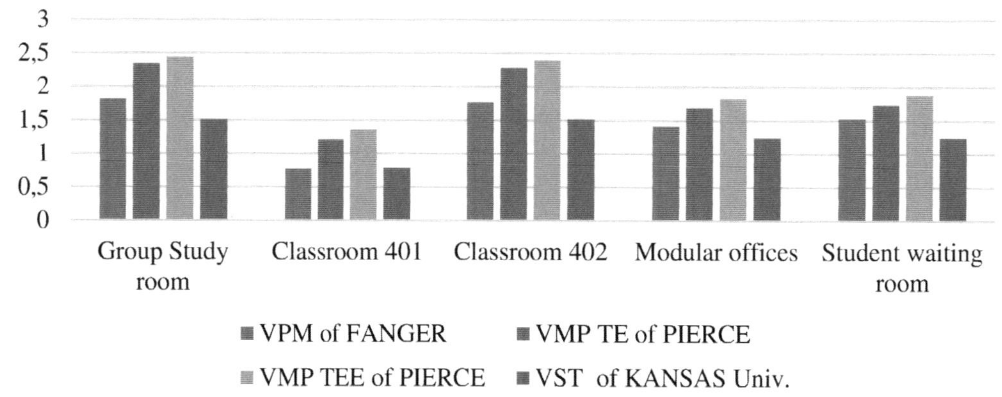

Figure 3. Thermal comfort indices for the most representative critical areas of EEB.

3.2. *Internal heat gains*
An analysis of the results of internal heat gains was conducted to identify the possible causes of thermal discomfort inside the building.

The heat through the glazing and heat associated with occupancy were the most significant internal sources of heat according to the simulations. These represent on average 68% and 17% of total internal gains, respectively. This makes them the main causes of thermal discomfort in areas such as Study group room and Classroom 402. Computers represent another important source of heating in areas, such as in Study room and Modular office with 14% and 26%, respectively.

3.3. *Effects of solar tubes on daylighting*
The effect of solar tubes on daylighting was analyzed in two critical classrooms located on the fourth floor of the building. The average daily value of the level of daylighting was established, taking into account partly cloudy sky conditions. An increase of 12.52% in the level of daylighting was found in the Classroom 402 due to solar tubes. In the case of the Classroom 401, the increase was around 40% due to a 100% increase in the amount of solar tubes present in this area with respect to Classroom 402. Another result showed that 58% of the areas susceptible to glare are areas of movement of persons.

3.4. *Effects of green roof on the building energy performance*
In order to demonstrate the effect of green roof on the building energy performance, the modular offices zone (5th floor) was selected. This area has the largest area of green roof exposed to solar radiation. According to the simulations, the green roof reduces approximately a 31% the value of internal heat gains associated to the cover with respect to a situation without green roof. The existing green roof reduces the annual energy consumption of the Classroom 401 near to 5%.

4. Conclusions
According to the process of modeling and simulation of EEB, DesignBuilder is a suitable energy simulation tool for analyzing buildings in the tropical environment. The simulation results show that

natural ventilation, solar radiation and occupation have direct influence on the energy performance of the building.

The results of thermal comfort showed that the most unfavorable environmental conditions were presented in the Group study room and Classroom 402. This behaviour is due to internal heating of the space by the action of solar radiation on the glazing, the heat radiated by the occupation and computers. Although most critical areas present conditions of thermal discomfort during most of the year, the natural ventilation minimizes the perception of those conditions.

Satisfactory levels of daylighting were found in every critical area, particularly those located on the upper floors. Solar tubes increase approximately 33% the level of daylighting. Of all the zones analysed, only 9% of them have glare conditions.

Green roof reduce a 31% of internal heat gains associated with the cover for administrative area of the building (5th floor); whilst, it produces near to 5% of energy savings in the Classroom 401.

The simulated energy consumption was approximately 50 000 kWh per year; this differs from the measured value at only 4 %.

5. References

[1] Peréz L, Ortiz J and Pout C, 2007, A review on buildings energy consumption information, *Energy Build,*. vol. 40, pp. 394–398.

[2] Mustafaraj G, Marini D, Costa A and Keane M, Oct. 2014, Model calibration for building energy efficiency simulation, *Appl. Energy*, vol. 130, pp. 72–85.

[3] El Sherif S K, April 2012, The Impact of Overhangs and Side-fins on Building Thermal Comfort , Visual Comfort and Energy Consumption in the Tropics.

[4] Jahnkassim P S, 2005, Energy strategies in the tropical context – the use of simulation tools in informing urban bioclimatic design in a malaysian highrise development – description and environmental study, vol. 2005, no. September, pp. 27–29.

[5] Oropeza I and Alberg P, 2014, Energy saving potential of utilizing natural ventilation under warm conditions – A case study of Mexico, *Appl. Energy*, vol. 130, pp. 20–32.

[6] Prianto E and Depecker P, 2003, Optimization of architectural design elements in tropical humid region with thermal comfort approach, *Energy Build.*, vol. 35, pp. 273–280.

[7] Mallick F H, 1996, Therman comfort and building design in the tropical climates, *Energy Build.*, vol. 23, pp. 161–167.

[8] Jamaludin N, Khamidi M F, Abdul S N and Klufallah M M, Aug. 2014, Indoor Thermal Environment in Tropical Climate Residential Building, *E3S Web Conf.*, vol. 3, p. 01026.

[9] DesignBuilder S, 2011, DesignBuilder EnergyPlus Simulation Documentation.

Stepwise drying of medicinal plants as alternative to reduce time and energy processing

S P Cuervo-Andrade[1] and O Hensel[2]

[1]Universidad Pontificia Bolivariana seccional Bucaramanga-Department of Mechanical Engineering. km 7 Vía Piedecuesta- Santander, Colombia
[2]University of Kassel- Department of Agricultural Engineering. Nordbahnhofstr. 1a. D – 37213 Witzenhausen, Germany.
E-mail: sandra.cuervo@upb.edu.co

Abstract. The objective of drying medicinal plants is to extend the shelf life and conserving the fresh characteristics. This is achieved by reducing the water activity (aw) of the product to a value which will inhibit the growth and development of pathogenic and spoilage microorganisms, significantly reducing enzyme activity and the rate at which undesirable chemical reactions occur. The technical drying process requires an enormous amount of thermal and electrical energy. An improvement in the quality of the product to be dried and at the same time a decrease in the drying cost and time are achieved through the utilization of a controlled conventional drying method, which is based on a good utilization of the renewable energy or looking for other alternatives which achieve lower processing times without sacrificing the final product quality. In this work the method of stepwise drying of medicinal plants is presented as an alternative to the conventional drying that uses a constant temperature during the whole process. The objective of stepwise drying is the decrease of drying time and reduction in energy consumption. In this process, apart from observing the effects on decreases the effective drying process time and energy, the influence of the different combinations of drying phases on several characteristics of the product are considered. The tests were carried out with Melissa officinalis L. variety citronella, sowed in greenhouse. For the stepwise drying process different combinations of initial and final temperature, 40/50°C, are evaluated, with different transition points associated to different moisture contents (20, 30, 40% and 50%) of the product during the process. Final quality of dried foods is another important issue in food drying. Drying process has effect in quality attributes drying products. This study was determining the color changes and essential oil loses by reference the measurement of the color and essential oil content of the fresh product was used. Drying curves were obtained to observe the dynamics of the process for different combinations of temperature and points of change, corresponding to different conditions of moisture content of the product.

1. Introduction

Lemon Balm (*Melissa officinalis* L.) is important due to its medicinal properties and its use as tea. Its essential oil is required mainly in the pharmaceutical, food and cosmetic industries. The properties of the plant extract include sedative, relaxing, antibacterial, antiviral, and antispasmodic effects [1,2].

The main components of the essential oil are citral, citronellal and linalool. By means of gas chromatography studies of the essential oil, 70 components are known [3,4,5]. Lemon Balm has low essential oil content (among 0.05 and 0.12 % vol.) [3].

Drying food, agricultural, medicinal plants and biological active materials is an important operation in agricultural, food, [6] and pharmaceutical industries to improve the preservation properties of the product, to reduce transportation costs in order to make the utilization of the products easier and efficient. [7,8]. The reduction of water content and the increase of the temperature of the solid during drying can lead to destruction or decrease of product quality [10]. Structural and physical modifications can take place, as for example the coloring, crust formation, decrease of organoleptic quality, inactivation of bacteria and enzymes, loss of nutrients and aroma. Additionally, drying processes are highly energy consuming, representing about 20-25% of the energy used in food processing industry [9,11]. The objectives of an energy efficient process and the highest possible product quality are conflicting in many cases. Optimization allows the explicit formulation of energy and quality objectives, and will therefore have a large potential in enhancing the operation of drying processes [10].

In the optimal control of drying of biological material, the quality of the material is the most important aspect. Since dried products are used in various situations and in various forms, it is obvious that the corresponding quality attributes are different. The quality attributes are linked to measurable quality indicators. Among many factors affecting the quality indicators, the most important are temperature and moisture content. As during the drying process, gradients of moisture content and temperature arise inside the material, the temperature and moisture distributions characterize the quality indicator rather than the average moisture content and average material temperature [12]. For instance, in the case of medicinal plants preservation, prevention of microbiological deterioration or the growth of organisms that could give rise to product poisoning is achieved by not exceeding a safe moisture content which is based on the local moisture concentration. Consequently, the quality of biological material is based on the local evolution of moisture and temperature profiles in the materials, irrespective of the defined quality indicator [11,13].

Several drying techniques have been proposed in order to rationalize the use of energy, as well as to reduce the negative effects of quality during the drying process. These effects could be cracks, fissures, non-enzymatic browning of fruits and vegetables, and lose of essential oils. One of these techniques is the intermittent drying [9,13]. This process alternates drying periods with rest or relaxation periods. The rest period (nondrying, tempering) allows homogenization of moisture through cooling. This applies to intermittent drying only in period when the drying rate is decreasing.

This simplified classification of the intermittency types is based on a literature reviews. The values of the focused used variables, like temperature, humidity, velocity and pressure, can be changed. The intermittency of these variables should not be chosen arbitrarily, rather it has to be selected based on the physics involved in the drying method, keeping in mind optimum energy efficiency and product quality. The development of new drying strategies for medicinal plants is required, so that it allows reaching the required quality characteristics and at the same time to have short drying times and low energy consumption. The strategies should consider an optimal balance between required quality and production costs.

The objectives of this research were to compare the conventional drying with the proposed stepwise drying, for lemon balm, considering drying time, energy consumption, color change of leaves and essential oil content, and additionally to determine the optimum change point of stepwise drying in terms of energy saving and reduction of drying time considering at the same time obtaining the quality characteristics.

2.Materials and Methods

The used sample of Lemon Balm (*Melissa officinalis* L.), variety citronella, was planted in a greenhouse in Witzenhausen (Germany). For this research two methods of drying were considered. The standard drying, is conducted at only one temperature (40°C) and stepwise drying using two temperatures.

The drying experiments were conducted using a cabinet dryer developed by the company INNOTECH in close cooperation with the Hohenheim University. Due to the construction of the dryer and application of the over current principle a low airflow resistance is reached within the equipment. This results in a low energy requirement of the fan as well as the possibility to use cross flow fans. The application of this type of fans leads to a uniform air distribution in the drying chamber. In combination with the registered profiled layout of the trays there is no need to shift single trays during the drying process to reach the desired uniform drying rate.

The experimental data for the drying of Lemon Balm in a convective tray dryer was obtained in the first step at a low temperature (40°C) and in the final step with high temperature (50°C) and having different changes points corresponding to different moisture content of the product as shown in Figure 1 the amount of water removed during the drying process was determined by periodic weighting. A digital weighing apparatus (with resolution of ±0.001g) measures the mass loss of the product during the drying process. During each drying experiment, the weight of the products on the tray was measured. These measurements were undertaken every 10 min the first hour of the experiment and then at every 60 minutes until products reach the humidity level of 10%. This process had to be done fast in order to avoid the re-absorption of water from the environment [10].

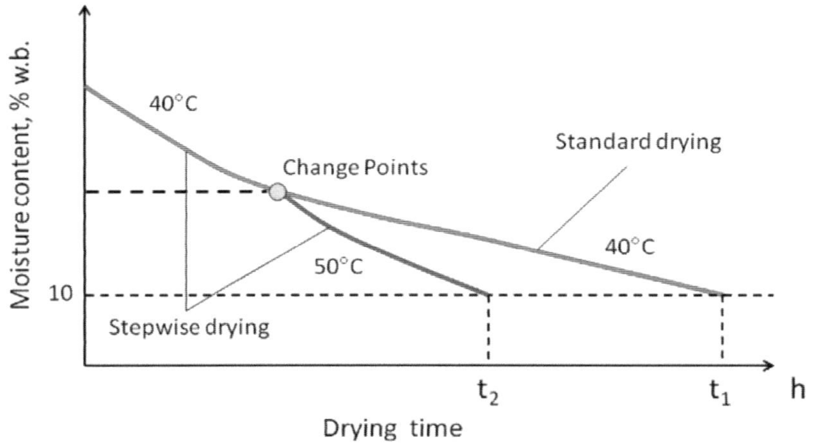

Figure 1: Drying methods considered in the study: standard drying and stepwise drying

The different change points of temperature, considered in the research, are related to the moisture of product in a wet basis. The first change point was taken at 50%, then to 40, 30 and finally to 20% since the process ends when the product is at 10% humidity.

During the drying process the moisture content loss of the product was measured by monitoring the weight until a moisture content of 10% is reached. Parallel, the consumed energy during the drying process was measured. The essential oil content and color change were determined before (fresh product) and then after (dried product) the drying process to be used as comparison for the results.

2.1 Moisture content

To determine the moisture content of Lemon Balm before and after drying, the oven method was used. In this method the sample is placed inside an oven at 103◦C (±2 ◦C) for 24 hours and the loss of weight is registered in order to determine the moisture content of the sample (Association of Official Analytical Chemists, 1990)

2.2 Color
The method used for measuring the color of leaves of Lemon Balm, before and after reached a 10% moisture content value with the drying process, was the same for all the experiments. For measurements, a Chroma Meter Minolta CR-400© was used. The chromameter was calibrated before the measurement by using a white ceramic plate as reference. Additionally a sample size of five leaves selected randomly was considered. Furthermore for measuring the color of fresh product the samples were taken the same day of the experiment. The evaluation of color was based on chromaticity values in the CIELAB color space. In the CIELAB system the L-axis is defined in a range between (0) and (100) which represents darkness (0) to lightness (100). The a-axis describes the green to red intensity on a negative to positive scale, while the b-axis is used in the same manner to describe the blue to yellow intensity.

The color is one of the most important factors to determine the quality of consumers and behind consumers of buying. Effect of stepwise drying methods have been investigated. For color measurements, leaves were separated from branch. To obtain the color change the norm DIN 6174 [15] was used. For the determination of color difference, the values of the reference sample were compared with each of the measurements of the samples of dried product. Later on the average of the color differences was calculated.

2.3 Essential oils
For the determination of the content of essential oil the method described in DAB 10 [17] was used. This method uses steam distillation with a special distillation device. To have a basis for comparison, this methodology is applied using 100g of fresh leaves in a liter of distilled water and 20g dry leaves was taken in 200ml of distilled water. The essential oil content in the samples is reported for 100g of dry matter for comparison there between. Losses of essential oil were calculated in percentage based on the difference of the essential oil content between the fresh leaves and samples taken from the dried material.

2.4 Energy
The energy consumption was measured by a Voltcraft Energy-check 3000 electrical monitor which was connected to the drying during each trial. Solids content was determined after each experiment by the oven drying method.

3. Results and analysis
Table 1. Shows the mass change with regard of time during the drying process on each of the change points.

Table 1. Mass change during drying time

Time [min]	W50% [g]	Time [min]	W-40% [g]	Time [min]	W-30% [g]	Time [min]	W-20% [g]	Time [min]	W-10% [g]
0	300.10	0	300.195	0	300.036	0	300.106	0	300.106
60	238.44	60	247.182	60	238.386	60	238.442	60	238.442
120	213.02	120	213.084	120	206.500	120	213.021	120	213.021
180	191.01	180	193.793	180	196.491	180	191.015	180	191.015
240	172.14	240	176.666	240	167.851	240	172.142	240	172.142
300	160.59	300	160.638	300	156.848	300	160.591	300	160.591
360	144.48	360	147.660	360	144.446	360	144.480	360	144.480
420	132.16	420	132.204	420	134.379	420	131.804	420	131.804
430	123.09	480	125.398	480	122.793	480	122.822	480	122.822
440	116.28	540	116.319	540	115.735	540	115.762	540	115.762
450	108.11	600	113.447	600	112.331	600	112.358	600	112.358
460	99.786	610	102.431	660	104.263	660	102.712	660	102.712
470	94.877	615	98.346	720	98.633	720	98.657	720	98.657
480	91.660	617	92.998	725	88.175	780	91.863	780	89.446
490	85.432	622	85.657	730	82.260	820	86.470	885	84.325
500	83.122	627	83.699	734	75.806	825	85.754	955	81.955
510	80.565	632	80.110			830	83.671	1020	80.012
525	78.332	637	78.954			835	79.877	1080	78.645
537	76.110	642	75.984			839	75.655	1140	76.227

In Figure 2, the drying curves corresponding to standard drying, for drying air temperature of 40°C, and stepwise drying combination of 40°C/50°C, considering different change points, are presented.

A considerable decrease on drying time compare to standard drying is observed. As seen in Figure 2, for all temperature changing points the drying time decreased compared to the standard to up until a maximum of 51% of time at 50% of humidity. The minimum decrease of drying time was 26% at humidity at 20%.

To analyze the change on color and essential oil content, the values for standard drying were considered as reference. In Figure 3, it can be observed that when the change point is higher, the change on color increases and the essential oil content decreases.

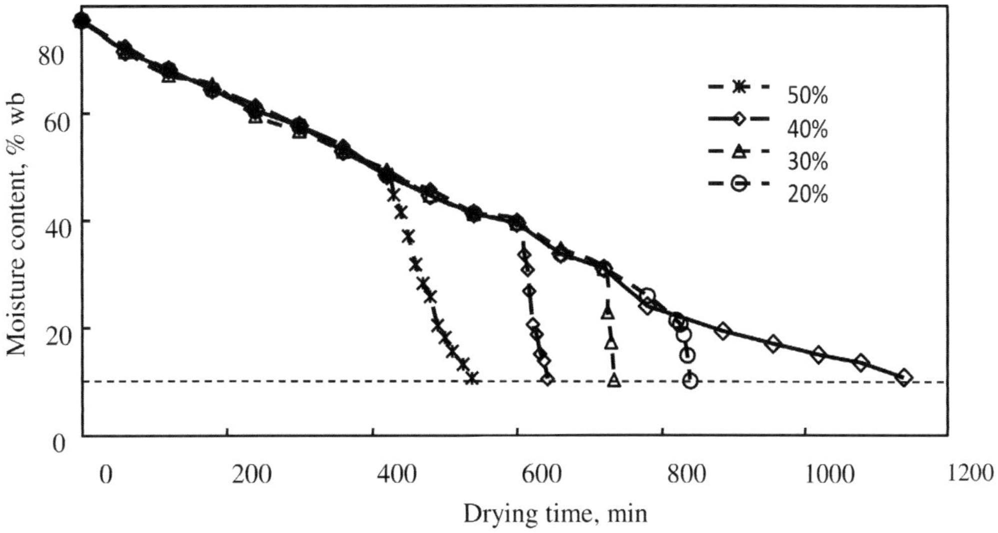

Figure 2: Considered drying methods: standard 40°C and stepwise 40°C/50°C

For standard drying, the obtained essential oil content was 0.17 ml/100gr dm (dried matter) and the color change was ΔE= 12. In contrast for stepwise drying with change points of 50%, the essential oil content was 0.06 ml/100gr dm, which corresponds to the largest loss of essential oil observed (near 65% of the reference value). Regarding the color change, for stepwise drying with change point 50%, ΔE= 22 was obtained, which was the largest color change observed.

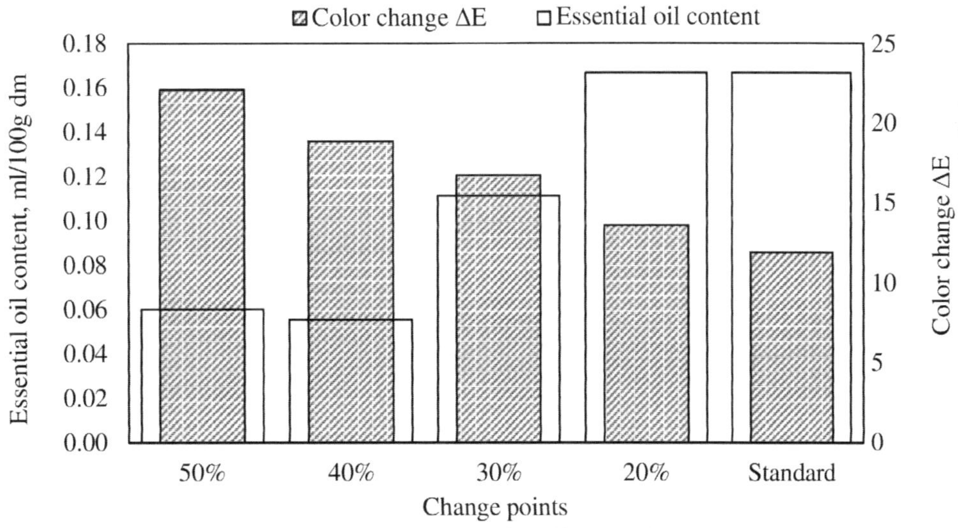

Figure 3: Color change and essentials oil content for different change points of the stepwise drying (40°C/50°C) compared with standard drying (40°C).

To analyze the consumed energy and drying time, the values for standard drying were considered as reference. In Figure 4 shows the behavior of drying time and consumed energy in the process for different change points of the stepwise drying compared with the standard drying. It can observed that if the change points increases, both the drying time and the consume energy decreases. For standard drying, the consumed energy was 4.25 kWh and the drying time was 1140 minutes. In comparison, for stepwise drying with change points of 50% the consumed energy was 3.07 kWh, which corresponds to a reduction of 28.5%. Regarding the drying time, for stepwise drying with change point 50% was 537 minutes, corresponding to a decrease of 53% in comparison with standard drying.

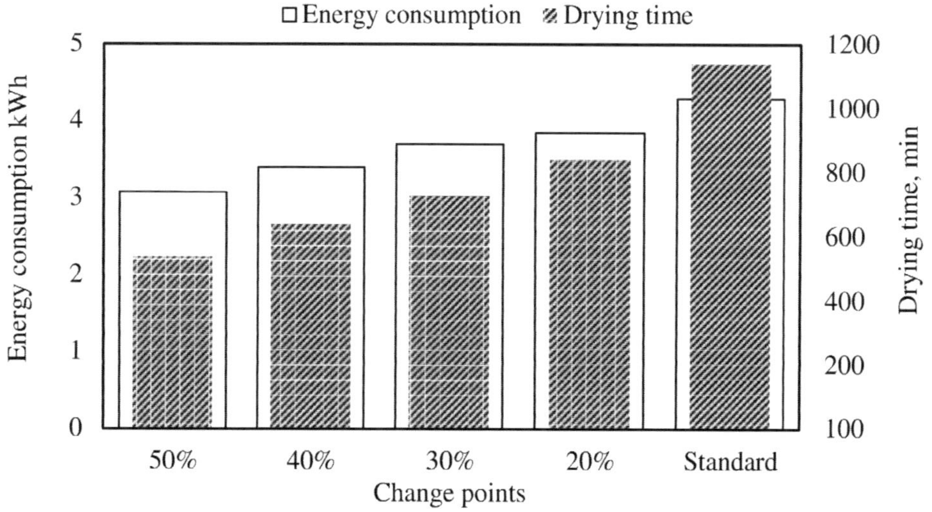

Figure 4: Drying time and consumed energy for different change points of the stepwise drying (40°C/50°C) compared with standard drying (40°C).

4. Conclusions

A decrease in drying time for all changes points in comparison to the standard drying was observed. The lowest energy consumption and shortest drying time was observed for change point 50%. A decrease of 50,6% in drying time can be reached together with a reduction of 28% in consumed energy compared with conventional drying at no changes in quality. But in terms of quality, the change point 50% shows high color change and the lowest content of essential oil, being this the least effective method.

The combination 40°/50°C, for change point 20% shows nearly no color change and the same essential oil content of standard drying, which makes it the most adequate combination in terms of quality. In terms of energy this combination reduces the energy consumption by 10% regarding the standard. The drying time was reduced by 28.5%.

References

[1] Ayanoglu F, Arslan M and Hatay A 2005 Effects of harvesting stages, harvesting hours and drying methods on essential oil content of Lemon balm in Eastern Mediterranean *International Journal of Botany* **1**(2) 138-142

[2] Heeger E 1980 Handbuch des Arznei-und Gewürzpflanzen Drogengewinnung VEB Deutscher Landwirtschaftsverlag

[3] Basta A, Tzakou O, Couladis M 2005 Composition of the leaves essential oil of Melissa officinalis L. S.I. from Greece Flavor and Fragrance Journal **20** 642-644

[4] Blum H and Lorenz J 2005 Ergebnisse der vergleichenden Prüfung von drei Sorten der Zitronenmelisse (Melissa officinalis L.) Zeitschrift für Arznei-und Gewürzpflanzen **10**(3) 133-139

[5] Bomme U 2001 Kulturanleitung für Zitronenmelisse. 4. Grundlegend überarbeitete. *Bayerische Landesanstalt für Bodenkultur und Pflanzenbau*

[6] Koyuncu T, Tosun I and Pinar Y 2007 Drying characteristics and heat energy requirement of cornelian cherry fruits (*cornus mas* L.) *Journal of Food Engineering* **78** (2) 735-739

[7] Chua K J, Mujumdar A S, Hawlader MNA, Chou SK and Ho J C 2001b Batch drying of banana pieces–effect of stepwise change in drying air temperature on drying kinetics and product colour *Food Research International* **34** (8) 721–731

[8] Kowalski SJ and Pawłowski A 2011a Energy consumption and quality aspect by intermittent drying *Chemical Engineering and Processing: Process Intensification* **50** (4) 384– 390

[9] Ramallo LA, Lovera NN., Schmalko, M.E., 2010 Effect of the application of intermittent drying on *Ilex paraguariensis* quality and drying kinetics *Journal of Food Engineering* **97** (2) 188–193

[10] Cuervo-Andrade S P 2011 Quality oriented drying of Lemon Balm (*Melissa officinalis* L.) *Forschungsbericht Agrartechnik des Fachausschusses Forschung und Lehre der Max-Eyth-Gesellschaft Agrartechnik im VDI (VDI-MEG)* (Diss. Witzenhausen)

[11] Kumar C, Karim MA and Mohammad UH 2014 Intermittent drying of food products: A critical review *Journal of Food Engineering* **121** 48-57

[12] Quirijins EJ 2006 Modelling and dynamic optimization of quality indicator profiles during drying *PhD. Thesis Wageningen University, Wageningen*, the Netherlands

[13] Barbosa de Lima A, Delgado J Neto S and Franco C 2016 Intermittent Drying: Fundamentals, Modeling and Applications *Drying and Energy Technologies, Advanced Structured Materials* **63** 19-41

[14] Association of Official Analytical Chemists AOAC 1990 Official methods of analysis: Moisture Content in plants 1949

[15] DIN 6174. 1979. Farbmetrische Bestimmung von Farbabständen bei Körperfarben nach der CIELAB-Formel

[16] Özgüeven M and Andana W 1999 Ertrags-und Qualitätseigenschaften von Melisse-Wildsammlungen aus der Südost-Turkey *Zeitschrift für Arznei-und Gewürzpflanzen* **4**(1) 39-43

[17] Deutsches Arzneibuch DAB 10 Komentar 1986 Wissenschaftliche Verlagsgesellschaft mbh Stuttgart

Detection and characterization of defects in moving parts of wind turbines

E Forero[1] D Tibaduiza[1] M Anaya[1] R Castro[1]
[1] Facultad de Ingeniería Electrónica, Universidad Santo Tomás –Colombia, Bogotá

E-mail: {edwinforero , diegotibaduiza, maribelanaya, rafaelcastro }@usantotomas.edu.co

Abstract. The detection, localization and characterization of defects in a material or a part that conform a structure is possible by using the transmission and reception of ultrasonic signals. Different strategies are used to achieve extract information from the part under evaluation. For this, it is then possible to use a distributed sensors arrays on the surface of the material and using scanning techniques such as are A-scan or B-scan, where it is possible to increase the level of detail regarding location, orientation and size of defects found, according to the strategy used. However, the systems and inspection techniques are often limited by the geometries and access to different types of structures. Due to these reasons, the acquisition of the returned signals, for identification and attenuation time, can suppress valuable information for accurate characterization of imperfections found in shape and location. In this paper, the use of spectral analysis of the collected signals is proposed as a tool for detection and characterization of defects in a structure. This analysis allows to determining the power distribution in a frequency range. This methodology is useful in non-destructive evaluation when it is not possible to have full access to the structure under inspection. In this case it is applied on a wind turbine operating to make the study of different echoes captured according to the geometry of the part and comparing said conducting analysis with previously established patterns of shapes, orientations, and sizes of defects found.

Introduction

Inspection of materials and component parts of structures allow a reliable criterion in certifying their quality. The inspection process is performed by capturing data using different techniques; for instance, the emission and reception of waves over the structure by piezoelectric active sensors [1], and Fiber Bragg Gratings using lamb waves [2], [3]. The piezoelectric technique is performed with transducers whose resonant frequencies are between the orders of kHz. Also is important to perform suitable signal capturing by the acquisition system, and this process involves the develop of specific data treatment modules for subsequent use of different signal processing techniques in order to identify the existence and nature of the found defects, so the work of the inspector provided along with methodologies attempt to find adequate precision.

[1] edwinforero@usantotomas.edu.co

In this work operation frequency spectral analysis is presented as in [4] and [5] to distinguish defects in the surface of a material, but with some changes such as data organization, pre-processing and data analysis. The methodology is tested using damages which are produced over the aluminium alloy structure, so the technique is destructive due to characterize the real damage nature and relational behaviour at the frequency spectral domain. This paper continues with a description of the experiment, then a frequency analysis is presented and finally the conclusions of the work are written.

Physical Principle

In solid materials, molecules can be forced to vibrate in different amplitudes and frequencies by applying diverse kinds of sound waves at any direction. The interaction between these mechanical waves and moving surfaces allow to extract information of damages and defects. Waves traveling on such surfaces are called lamb waves, they are generated by a piezoelectric transducer added to the surface under study. Lamb waves can vibrate in different modes, but the most common are asymmetric or symmetric, in this sense, the wave frequency characteristic can be an indication of the changes in the structures such as different damage shapes and surface boundaries.

Description of the experiment

For the experimental setup, an aluminium alloy structure was instrumented with two piezoelectric sensors attached permanently to the extremes of one of its faces (figure 1a). Both piezoelectrics, act as actuator/sensors in different actuation phases of the experiment, this configuration is used to apply the input and for the acquisition of the signals [13]. The input signals are generated by means of a burst signal with 8V of amplitude (figure 1b). The data are collected with an HS4 TiePie oscilloscope at a 2 MHz sample rate, and organized by actuation phases as in Tibaduiza 2013 [5]. The input signal generation and acquisition program were developed in MATLAB.

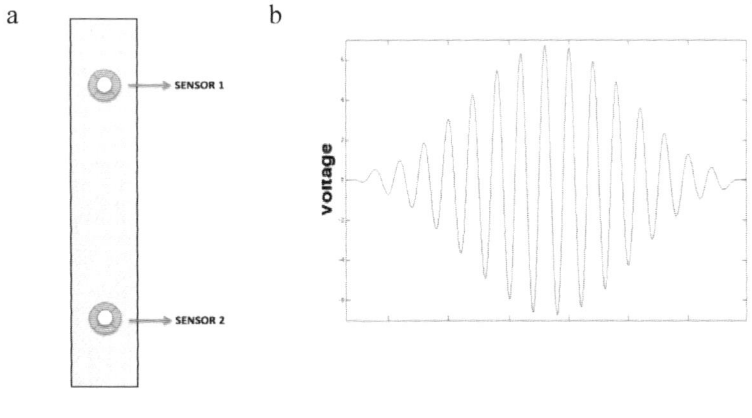

Figure 1. a) Shows the arrangement of piezoelectric transducers permanently attached to the aluminum structure on one of its faces. b) input signal, burst 10 kHz in hamming window

To carry out the work, six different damages were applied to the structure. The damages included are: one surface circular damage with 3.5mm radius (figure 2b), one round hole with 4.7 radius (figure 2c), one round hole with 6.2 radius (figure 2d), one equilateral triangle hole with 12mm of side (figure 2e), one right trapezoidal hole with base and side of 12mm (figure 2f) and one square hole 12mm side (figure 2g).

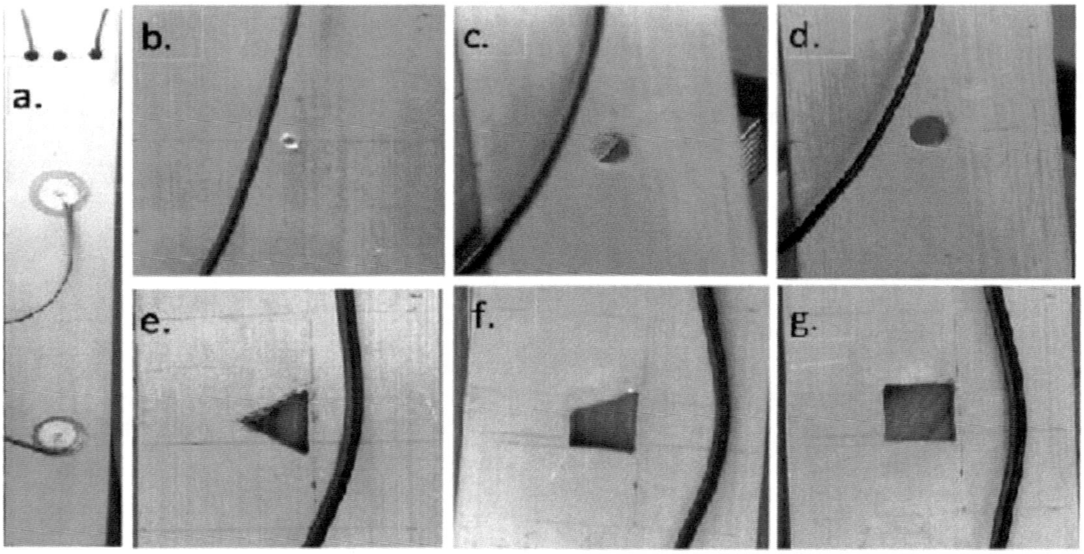

Figure 2. a) Healthy structure. The evaluated defects: b) Surface circular, c) round hole with 4.7 radius. d) Round hole with 6.2 radius. e) Equilateral triangle. f) Right trapezoidal. And g) Square hole.

In the acquisition process there were performed in total one hundred forty measurements related to twenty data obtained for each one of the six different geometries and twenty for the undamaged case. The type of acquired signals are shown in figure 3.

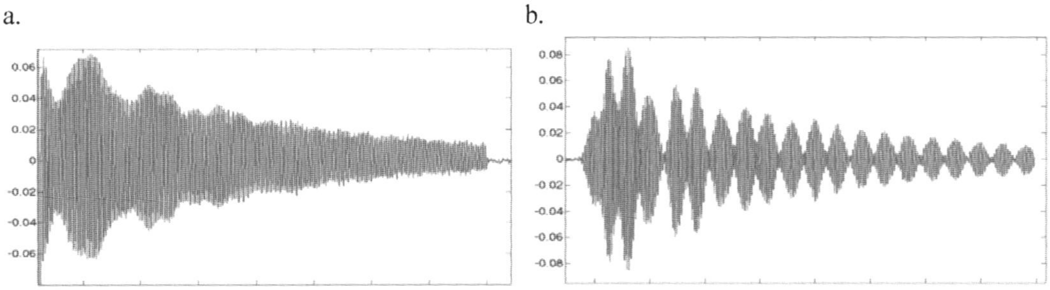

Figure 3. The acquired data: a) Return wave for the actuation sensor. b) Wave of signal by the received sensor.

For all the damaged cases, the data were evaluated in the frequency domain considering each actuation phase in order to obtain the central frequency.

Frequency Analysis

The acquired data were organized in a new dataset containing all data from the undamaged and damaged structure, after the Discrete Fourier Transform (DFT) is applied. For all the damage cases the data were evaluated in the frequency domain and the results (power by phase [dB]) are included in the Figure 4.

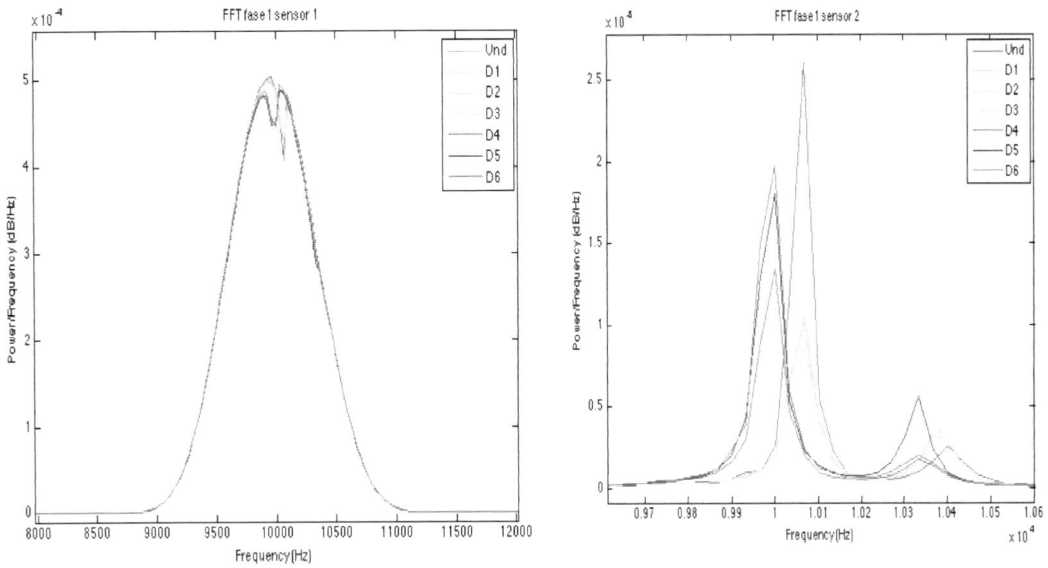

Figure 4. Figure with short caption (caption centred) of the healthy structure and the evaluated defects central frequency power for the phase 1; sensor 1 actuation and sensor 2 reception; D1- Undamaged structure. D1-Surface circular damage with 3.5mm radius. D2- Round hole with 4.7 radius. D3-Round hole with 6.2 radius. D4-Equilateral triangle hole with 12mm of side. D5- Right trapezoidal hole with base and side of 12mm. D6-Square hole 12mm side.

The obtained results (Figure 4) in the frequency domain show an important feature, this is the emergence of new frequency components, frequencies rounding the 10 KHz with different amplitudes in the spectrum, which will be evident in the defects distinction. In the actuating sensor, damages related to the equilateral triangle, the right trapezoidal and the square holes, shows in the spectrum different shape than for the round surface and holes damages, these lasts are more close in shape for the undamaged case, a relate behavior is observed for the receiving sensor spectrum, but in this case there is a shift of the frequency bands, showing a very different central frequency between round damages and the others. For the actuating spectrum, the main phenomena related to the frequency is the return wave, so it can be said that round damages produce a change in the magnitude but not in frequency responses. For the receiving sensor this change of central frequency is mainly related to the shape of the damage, the circular defects seem to be closer in frequency to the data from the undamaged structure presenting only differences in the magnitude, while the planar face damages have a central frequency component which is below the undamaged state, in the same way, there are different magnitudes. A graphic analysis of the responses in the frequency domain obtained by the discrete Fourier transform (DFT) for each of the events evaluated is shown in the table 1 and the magnitudes in the central frequencies for each case are included in the table 2.

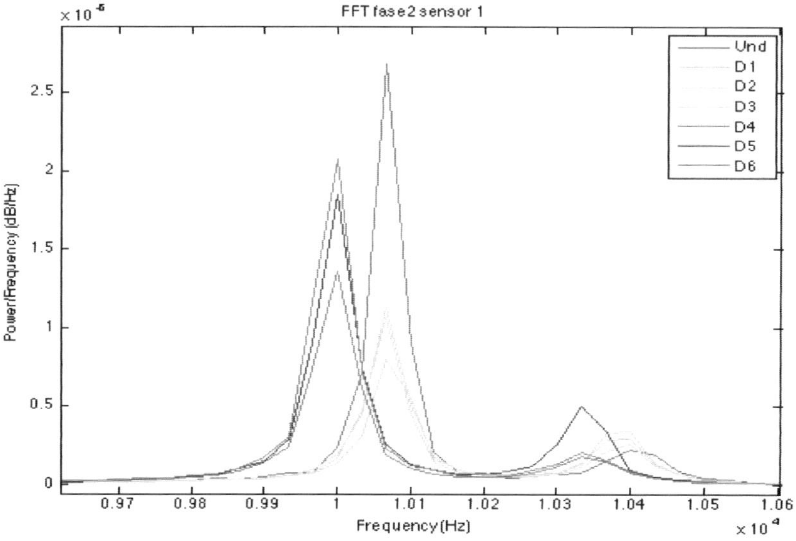

Figure 5. Figure with short caption (caption centred) of the healthy structure and the evaluated defects central frequency power: D1- Undamaged structure. D1-Surface circular damage with 3.5mm radius. D2- Round hole with 4.7 radius. D3-Round hole with 6.2 radius. D4-Equilateral triangle hole with 12mm of side. D5- Right trapezoidal hole with base and side of 12mm. D6-Square hole 12mm side.

A result from the previous plot (Figure 5), is that for the planar-faced damages, the central frequency magnitude in the receiving sensor seems to be increased with the damage size. In the same way there are some shifts in the central frequency and the amplitude by each actuation phase.

Table 1. Central frequency (Hz).

	Central frequency(Hz)			
	phase 1		*phase 2*	
	Sensor 1	*Sensor 2*	*Sensor 1*	*Sensor2*
Undamaged	10000,167	10067,058	10000,167	10033,612
Superficial and circular, radius 3,5 mm	9966,7217	10067,058	10100,503	9966,7217
Circular, radius 4,7 mm	9966,7217	10067,058	10100,503	9966,7217
Circular, radius 6,2 mm	9966,7217	10067,058	10100,503	9966,7217
Equilateral triangle, side 12mm	10067,058	10000,167	10033,612	9966,7217
Square, side 12 mm	10067,058	10000,167	10033,612	9966,7217
Right trapezoid, base and side 12mm	10067,058	10000,167	10033,612	9966,7217

Table 2. Power in the central Frequency (dB).

	Power in the central Frequency (dB)			
	phase 1		*phase 2*	
	Sensor 1	*Sensor 2*	*Sensor 1*	*Sensor2*
Undamaged	0,000503073	0,000494021	1,9677968E-06	2,0702139E-06
Superficial and circular, radius 3,5 mm	0,000497978	1,0590177E-06	1,1297371E-06	0,000518213
Circular, radius 4,7 mm	0,000496072	1,0113313E-06	1,0584284E-06	0,000517697
Circular, radius 6,2 mm	0,000499121	8,6029257E-07	7,9817715E-07	0,000518741
Equilateral triangle, side 12mm	0,000486173	1,3389920E-06	1,3606202E-06	0,000509451
Right trapezoid, base and side 12mm	0,000494021	1,9677968E-06	2,0702139E-06	0,00051361
Square, side 12 mm	0,000488466	1,8030465E-06	1,8498930E-06	0,000513379

Conclusions

The methodology herein presented, which is based on a piezoelectric transducer approach makes use of transducers which are working as sensor and actuators working in different actuation phases provides a good system to produce mechanical waves over materials. From the information acquired by frequency analysis we can see that both the receiving and the actuation sensor data provides a good differentiation between planar-faced damages and round damages with little shift over the frequency band. Round damages presented magnitude changes while conserving frequency similarities, finally in the case of the planar-faced damages, frequency changes and magnitude proportional to damage size were obtained.

Acknowledgments
This work is supported by Universidad Santo Tomas through through Grant FODEIN 2015 and 2016, project codes 8410083003- 036 and FODEIN 1608303-017.

References
[1] Anaya M, Tibaduiza DA, Forero E, Castro R and Pozo F 2015 *Proc. The Twentieth Symposium on Signal Processing, Images and Computer Vision-STSIVA2015 (Bogotá).*

[2] Edalati K, Kermani A, Seiedi M and Movafeghi M 2005 *Proc The 8th International Conference of the Slovenian Society for Non-Destructive Testing Application of Contemporaty Non-Destructive Testing in Engineering Portorz, Slovenia.*

[3] Haiyan Z, Xiuli S, Xuerui Q, Xiao L, Donghui L 2008 *Proc Microwave Conference, China-Japan Joint Pages: 805 - 808, DOI: 10.1109/CJMW.2008.4772550.*

[4] Chang N, Zheng T, Ren Y 2009 *Application of the Spectral Analysis in Ultrasonic Quantitative Testing of Bonding Defect of Composite Material, Education Technology and Computer Science*, ETCS '09. First International Workshop on. Volume: 3 Pages: 943 - 945, DOI: 10.1109/ETCS.2009.748

[5] Valença J, Gonçalves LMS, Júlio E 2013 *Damage assessment on concrete surfaces using multi-spectral image analysisOriginal Research Article*, Construction and Building Materials, Volume 40, March 2013, Pages 971-981.

[6] Tibaduiza D 2013 *Design and validation of a structural health monitoring system for aeronautical structures.* PhD thesis, Department of Applied Mathematics III, Universitat Politecnica de Catalunya.

[7] Ohtsua M and Watanabeb T 2002 *Stack imaging of spectral amplitudes based on impact-echo for flaw detection.* NDT & E International Volume 35, Issue 3, April 2002, Pages 189–196

[8] Chang N, Zheng T, Ren Y 2009 *Application of the Spectral Analysis in Ultrasonic Quantitative Testing of Bonding Defect of Composite Material in Education Technology and Computer Science*, 2009. ETCS '09. First International Workshop on, vol.3, pp.943-945.

[9] Bing L, Cao J, Xiao J, Zhang X and Wang H 2014 Proc. On Intelligent Control and Automation (WCICA), 2014 11th World Congress, pp.2854-2859, June 29 2014-July 4 2014.
[10] Demirli R and Saniie J *2009 Proc IEEE International Ultrasonic Symposium (IUS), pp.1522-1525, 20-23.*
[11] Yan Y, Hopgood JR, Steel R, Sboros V 2008 *Proc Ultrasonics Symposium, 2008. IUS 2008. IEEE, vol., no., pp.1382-1385.*
[12] Oruklu E, Aslan S, Saniie J 2009 *Applications of time-frequency distributions for ultrasonic flaw detection*, in Ultrasonics Symposium (IUS), 2009 IEEE International, pp.2000-2003.
[13] Anaya M, Tibaduiza D, and Pozo F 2015 *A Bioinspired Methodology Based on an Artificial Immune System for Damage Detection in Structural Health Monitoring*. Shock and Vibration, vol. 2015, Article ID 648097, 15 pages.

A Sensor Fault Detection Methodology applied to Piezoelectric Active Systems in Structural Health Monitoring Applications

D Tibaduiza [1], M Anaya[1,2], E Forero[1], R Castro[1], F Pozo[2].
[1] Faculty of Electronics Engineering, Universidad Santo Tomás, Kr 9 No. 51 – 11, Bogotá D.C., Colombia.
[2] Department of Mathematics, CoDAlab, Universitat Politecnica de Catalunya. Carrer Comte d'Urgell 187, Barcelona, Spain.

E-mail:{diegotibaduiza, maribelanaya, rafaelcastro, edwinforero }@usantotomas.edu.co, francesc.pozo@upc.edu

Abstract Damage detection is the basis of the damage identification task in Structural Health Monitoring. A good damage detection process can ensure the adequate work of a SHM System because allows to know early information about the presence of a damage in a structure under evaluation. However this process is based on the premise that all sensors are well installed and they are working properly, however, it is not true all the time. Problems such as debonding, cuts and the use of the sensors under different environmental and operational conditions result in changes in the vibrational response and a bad functioning in the SHM system. As a contribution to evaluate the state of the sensors in a SHM system, this paper describes a methodology for sensor fault detection in a piezoelectric active system. The methodology involves the use of PCA for multivariate analysis and some damage indices as pattern recognition technique and is tested in a blade from a wind turbine where different scenarios are evaluated including sensor cuts and debonding.

1. Introduction

Structural Health Monitoring (SHM) is an imperative need in the damage identification process. The use of a good SHM system allows to improve the inspection process and optimise the performance over uncertain conditions to ensure the security of the in-service structure in its continuous operation, through regular inspection for malfunctioning or damaging [1]. Damage diagnosis is often classified in different levels starting with the damage detection, where the existence of abnormalities gives information about the presence of damages to the SHM system [2]. In this sense, the reliability of the systems plays an essential role, and given the nature of some approaches, a fault in the sensor sub-system, results in a false-positive damage detection which produces that structure can be incorrectly classified [3]. Different sensors can be used in the inspection process, some of the most popular are the PZT sensors due to its versatility since they can work as sensors or as actuators producing and collecting different kind of

[1] diegotibaduiza@usantotomas.edu.co.

waves which are propagated trough the structure. Specifically, in data-driven algorithms which use pattern recognition techniques, the health of the piezoelectric transducers can be consider as a critical factor in the inspection process [4],[5],[6]. Common damages in PZT transducers include sensor debonding, piezoelectric fractures, bad connections which are produced even from the installation of the monitoring system or produced during life-time of the structure [7].

Previous work by the authors included the use of multivariate statistical methods to detect [8], localize [7] and classify [10] structural damages using PCA and some bio-inspired methodologies such as Self-Organizing Maps and Artificial Immune Systems [11]. In this paper, a fault detection methodology which is based on Principal Component Analysis and some damage indices is proposed and applied to an active piezoelectric system which acts in different actuation phases. To validate the methodology, sensor debonding, incremental cuts and crystal removals are studied, results show a good performance in the identification of the different scenarios. This paper is organized as follows: Section 2 includes a brief description of the Fault detection methodology and the experimental setup used for the validation, Section 3 presents the Results and finally some conclusions are discussed in last section.

2. Fault Detection Methodology and Experimental Setup

For the experimental setup a blade from a vertical axis wind generator ALEKO WGV75W was equipped with four piezoelectric sensors which were distributed over the two external faces, (see figure 1). The inspection process is performed by the PZTs working in different actuation phases as in [8], this active system uses each sensor acting as receiver and as actuator in the four different phases of the experiment. The excitation input was a 50 cycles and 100 KHz burst signal with 8 V of amplitude, data were collected at a 20 MHz sample rate, and organized by sensor actuation phases as in [10]. The signal generation and acquisition program were developed in MatLab®.

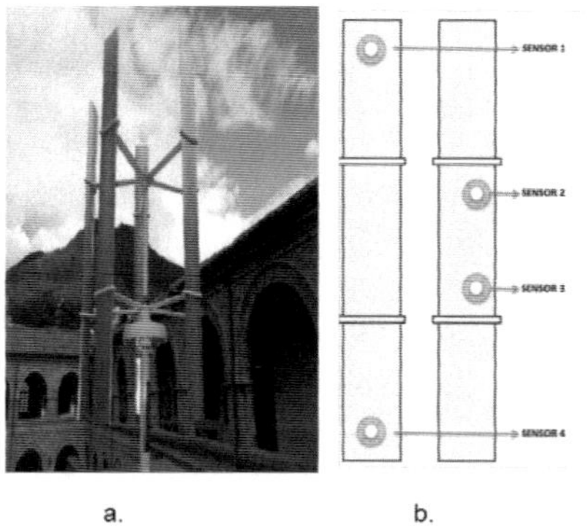

Figure 1. a) Vertical axis wind generator ALEKO WGV75W [9]. b) Distribution of the piezoelectric transducers.

The data from the structure in all the actuation phases are organized in different unfolded matrices when sensors are known as healthy and pre-processed using group-scaling [11]. These data are used to define the pattern by each actuation phases. After that, a PCA model is obtained by each actuation phase and the number of scores are defined according to the retained variance. Subsequently, with the structure and the sensors in different states are obtained, organized and pre-processed to be projected into the PCA-model by each actuation-phase. These projections are plotted and used to calculate the Q and T^2-

index [8][12]. The former is based on analysing the residual data matrix to represent the variability of the data projection within the residual subspace.

Eight damages were induced over one sensor, they are described in Table 1 and their appearance is shown in figure 2. Five of the eight damages are related to the piezoelectric crystal with cuts and removal of a certain area, the other three are related to the debonding of the metal plate of the sensor. Both, crystal removing and debonding damages are supposed to be critical in the actuation mode, since these can change the signal applied to the structure. In the same way, in the data acquisition, these damages can affect the data reliability since the mechanical transduction to the electrical field which is going to be corrupted for the cracks in the crystal and to the incomplete energy received (case of debonding).

Table 1. Sensor Induced Damages

Type	Used area	Brief description
Cut [a]	-	Cut to the middle of the crystal at 25% line
Cut [a]	-	Complete cut of the crystal at 25% line
Crystal removal	25%	Removal of the 25% area of the crystal
Crystal removal	50%	Removal of the 50% area of the crystal
Crystal removal	75%	Removal of the 75% area of the crystal
Sensor Debonding [b]	25%	Adherent in the 75% of the sensor's plate
Sensor Debonding [b]	50%	Adherent in the 50% of the sensor's plate
Sensor Debonding [b]	75%	Adherent in the 25% of the sensor's plate

[a] The damage doesn't compromise any area type of the sensor
[b] The damage compromises the sensor metallic plate area

Those damage were physically induced in the sensor 1, the other piezoelectric transducers and the wind blade structure were in a healthy state. Figure 3 presents a description of the percentage of debonding.

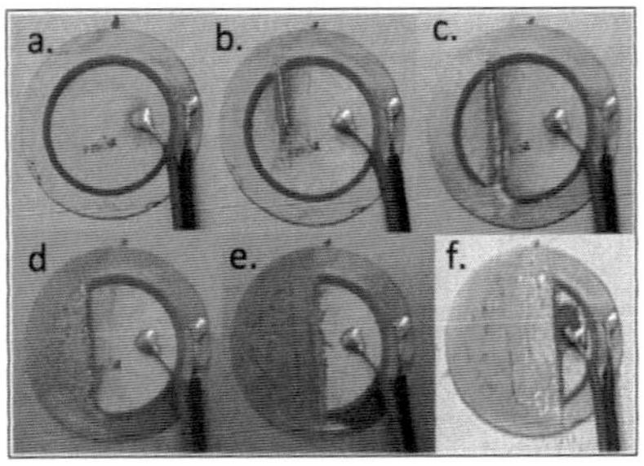

Figure 2. a) Undamaged sensor. b) Crystal middle cut at 25% line. c) Crystal complete cut at 25% line. d) Crystal 25% removal. e). Crystal 50% removal f) Crystal 75% removal.

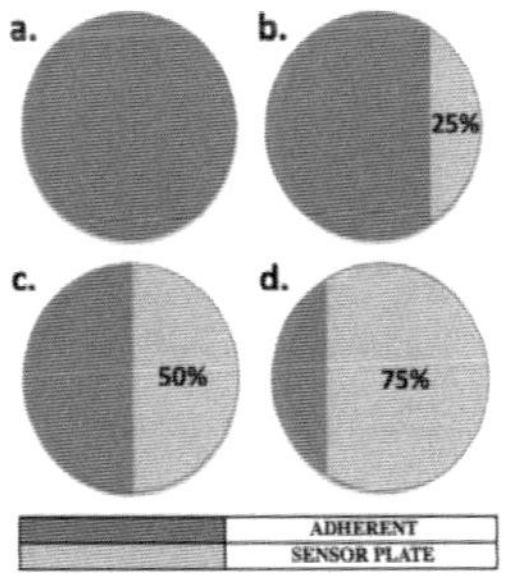

Figure 3. Debonding damage illustration, a) undamage. b) debonding 25%. c) debonding 50%. d) debonding 75%.

To show the influence of the damaged sensors in the damage detection process, 2 simulated damages which consists of magnets attached to the structure in order to change the mechanical impedance were introduced as in Figure 4.

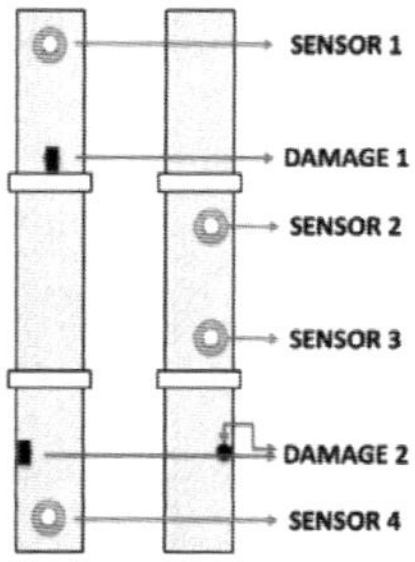

Figure 4. Two impedance damages over the wind blade structure, each damage is composed by magnets to represent a load in the structure, and which can be recognized by mechanical behaviour of the sensing devices.

3. Results

The use of the methodology previously introduced allows to obtain two different results. First, when the first two projections (scores) to the principal components are used, the score plots are obtained. Second, when the measurements T^2 and Q are used, the damage indices plots are obtained. These plots are calculated by each actuation phase and analysed with all the faults and structural conditions including to the healthy state.

As first outcome, Figure 5 shows the score and damage index plots using data from undamaged sensor, crystal cuts and removals were organized to evidence if each sensor fault could to be recognized by comparing with the baseline. By a simple comparison, can be concluded that in this case, score plot with just two components is not able to discern the presence of faults, however the use of the damage indices allow to determine the presence of all damages and some differences between them. Similar results are obtained in each actuation phase.

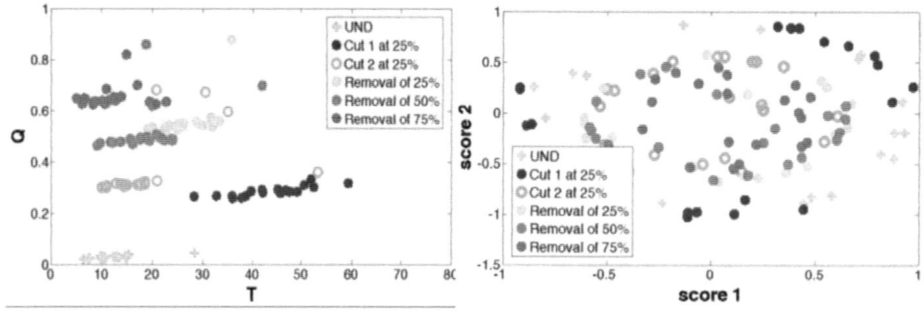

Figure 5. T-Q PCA indices and Scores 1-2 for the 5 damages induced over the piezoelectric crystal. (actuation phase 2)

Damages over the sensing system maybe critical, but in some cases it is not possible to replace or repair the sensors immediately. Considering this situation, the second test was performed by including simulated damages to the structure when one of the sensor is known as with a fault. Figure 6 shows the results in the use of the damage indices, as it is shown both damages can be separated from undamaged, however it is necessary to highlight that baselines were built with the sensor in the different fault conditions, this result allow to confirm that the pattern recognition approach allows to determine the presence of damages in spite of the faults in a sensor.

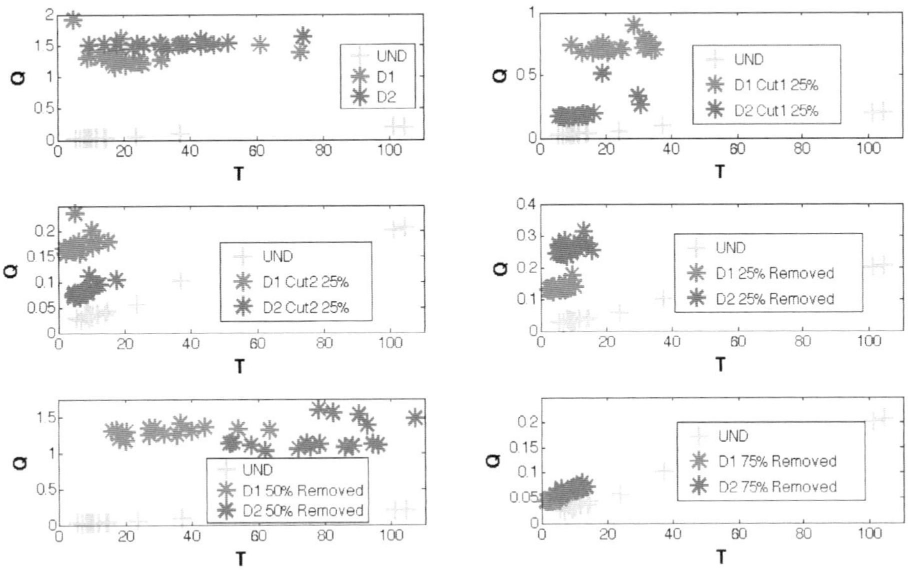

Figure 6. T-Q PCA indices for the 2 impedance damages over the structure in a. undamaged sensors b. fault the damages are measured from the 5 faults induced over the piezoelectric crystal (actuation phase 1)

Evaluating the rest of the actuation phases, similar results are obtained, although for the Phase 1 which includes the fault sensor at 75% crystal removal case, the damages seem to be close enough from each other, is the only case where they behave like this (figure 7).

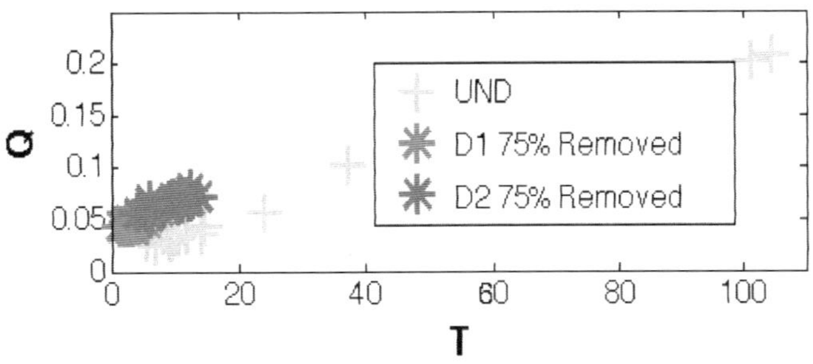

Figure 7. T-Q PCA indices for the 2 impedance damages over the structure, damages are measured from the 75% crystal over the piezoelectric.

As third experimental validation, the detection of debonding was performed. Figure 8 presents the results when an incremental debonding is studied in two actuation phases. As it is possible to observe, the undamaged state is separated from the different faults.

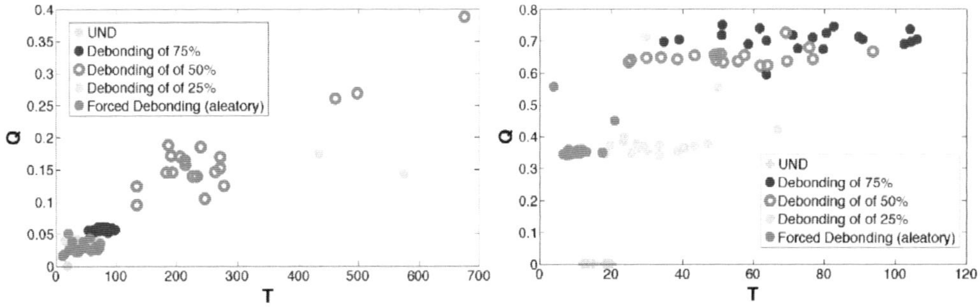

Figure 8. T-Q PCA indices for the debonding damages over the structure. For the actuation phases 1 and 2.

Conclusions

The PCA and Damage Indices based methodology allow to detect all the faults in spite of the differences between crystal cuts, crystal removal and debonding. Better results were obtained with the use of the damage indices, this is because two components are not sufficient to discriminate differences between the different scenarios. One of the disadvantage in the methodology is the big quantity of plots to analyse, in this case two plots by each actuation phase, however this paper shows that damage indices plots can be evaluated without the use of the score plots. This pattern recognition approach has the advantage of the baseline comparison, which can be related not only to the structure damage but also with the performance of the active sensor through its nominal behaviour. The methodology shows advantages for implementation in a SHM system, since it can be used as a first method to detect the current state of the sensors previously to the application of damage identification in a structure. Since multivariate analysis is needed, the pre-processing can be used to detect damages in the structure and faults in the sensors. In addition, since the methodology is based on a pattern recognition which can be applied to the analysis of data from sensors attached to the structures different to the PZTs used in this work.

Acknowledgments

This work is supported by Universidad Santo Tomas through Grant FODEIN 2015 and 2016, project codes 8410083003- 036 and FODEIN 1608303-017. In addition this work has been partially funded by the Spanish Ministry of Economy and Competitiveness through the research projects DPI2014-58427-C2-1-R and by the Catalonia Government through the research project 2014 SGR 859.

References

[1] Anaya M, Tibaduiza DA, Forero E, Castro R and Pozo F 2015 *Proc. The Twentieth Symposium on Signal Processing, Images and Computer Vision-STSIVA2015 (Bogotá).*

[2] Rytter A 1993 *Vibration Based Inspection of Civil Engineering Structures*. PhD thesis, Department of Building Technology and Structural Engineering. Aalborg University, Denmark, 1993.

[3] Fritzen CP, Kraemer P and Klinkov M 2010, *Proceedings of the IMAC-XXVIII, Jacksonville, Florida USA.*

[4] Sun Z, Rocha B, Wu KT, and Mrad N 2013 *A Methodological Review of Piezoelectric Based Acoustic Wave Generation and Detection Techniques for Structural Health Monitoring.* International Journal of Aerospace Engineering, vol. 2013, Article ID 928627, 22 pages, 2013. doi:10.1155/2013/928627

[5] Seong-Won Choi, Farinholt K, Taylor S, Light-Marquez A, and Park G 2014 *Damage Identification of Wind Turbine Blades Using Piezoelectric Transducers*, Shock and Vibration, vol. 2014, Article ID 430854, 9 pages

[6] dell'Isola F, Vestroni F, Vidoli S 2005 *Structural-Damage Detection by Distributed Piezoelectric Transducers and Tuned Electric Circuits*, Research in Nondestructive Evaluation, vol. 16, pp 101-118

[7] Tibaduiza DA, Mujica LE, Anaya M, Rodellar J 2011 *Combined and I indices based on principal component analysis for damage detection and localization, Proc. The 8th International Workshop on Structural Health Monitoring, Stanford, California (USA), 2011*

[8] Tibaduiza D 2013 *Design and validation of a structural health monitoring system for aeronautical structures.* PhD thesis, Department of Applied Mathematics III, Universitat Politecnica de Catalunya, 2013

[9] ALEKO GV75W 50W Nominal 75W Maximum 24V Residential Vertical Wind Power Generator. http://www.alekoproducts.com/ALEKO-WGV75W-50W-Nominal-75W-Maximum-24V-p/wgv75w24v-ap.htm. Consulted: November 2015.

[10] Tibaduiza DA, Mujica LE, Rodellar J 2013 *Damage classification in structural health monitoring using principal component analysis and self-organizing maps.* Structural Control and health Monitoring 20 (10), 1303-1316.

[11] Anaya M, Tibaduiza D, and Pozo F 2015 A Bioinspired Methodology Based on an Artificial Immune System for Damage Detection in Structural Health Monitoring. Shock and Vibration, vol. 2015, Article ID 648097, 15 pages.

[12] Alcala CF, Qin SJ 2009 *Proc. The 7th IFAC Symposium on Fault Detection, Supervision and Safety of Technical Process. Barcelona, Spain.*

Computational tool for simulation of power and refrigeration cycles

E Córdoba Tuta[1] and M Reyes Orozco[2]

[1]Universidad Pontificia Bolivariana, Facultad de Ingeniería Mecánica, km 7 vía Piedecuesta, Bucaramanga, Colombia
[2]Universidad Pontificia Bolivariana, Facultad de Ingeniería Mecánica, km 7 vía Piedecuesta, Bucaramanga, Colombia

E-mail: edwin.cordoba@upb.edu.co

Abstract. Small improvement in thermal efficiency of power cycles brings huge cost savings in the production of electricity, for that reason have a tool for simulation of power cycles allows modeling the optimal changes for a best performance. There is also a big boom in research Organic Rankine Cycle (ORC), which aims to get electricity at low power through cogeneration, in which the working fluid is usually a refrigerant. A tool to design the elements of an ORC cycle and the selection of the working fluid would be helpful, because sources of heat from cogeneration are very different and in each case would be a custom design. In this work the development of a multiplatform software for the simulation of power cycles and refrigeration, which was implemented in the C ++ language and includes a graphical interface which was developed using multiplatform environment Qt and runs on operating systems Windows and Linux. The tool allows the design of custom power cycles, selection the type of fluid (thermodynamic properties are calculated through CoolProp library), calculate the plant efficiency, identify the fractions of flow in each branch and finally generates a report very educational in pdf format via the LaTeX tool.

1. Introduction

This work aims to simulate refrigeration cycles and power cycles, specifically cycles *Rankine*. A computational tool was developed in the programming language C++ using the Qt cross-platform environment. It is also possible to simulate Organic Rankine Cycles (ORC) using a secondary fluid as the working fluid, Tchanche [1] made a detailed analysis of the different types of architectures for these cycles study. Programming in an object-oriented language has many advantages over a structured programming language, such as the work of Jeon [2] and Shirakawa [3], which use class inheritance for creating new models according to the needs. There are several developments in this kind of simulations in steady state and transient state, authors such as Fang [4] and diGenova [5] worked directly with commercial software such as Matlab© and interacted with other software to calculate thermodynamic properties, while authors Kim [6] and Akolekar [7] developed a custom software.

The first part consisted of the design of the models, starting with the modeling of the working fluid for which it was decided to use libraries *CoolProp*[8], modeling of the devices was based on existing in the literature concerning thermodynamic devices operating at steady flow models and then the cycle model was modeled with graph theory. Then it proceeds to solve algebraic

IOP Conf. Series: Materials Science and Engineering **138** (2016) 012017 doi:10.1088/1757-899X/138/1/012017

equations in the model obtained by solving a positive square matrix $N \times N$ using *Cholesky* decomposition. And finally the tool generates a report format *pdf*, and the results were validated with problems encountered in the literature.

2. Modeling

The model of the entire system comprising: modeling fluid; modeling each device and the cycle modeling, which are detailed below.

The calculation of the thermodynamic properties of fluids are made by libraries *CoolProp*[8]. CoolProp is an open source library written in `C++` and it implements pure and pseudo-pure fluid equations of state and transport properties for 114 components and humid air also. It can also interact with the program `REFPROP` [9] for calculating properties.

The modeling of the devices was made to conditions of steady-state flow and generally is defined by the energy balance equation 1 and the mass balance equation 2. These equations use the sign convention [10] for heat and work, which is widespread in the literature.

$$\sum_{out} \dot{m}h - \sum_{in} \dot{m}h = \dot{Q} - \dot{W} \tag{1}$$

where $\dot{m}$ is the mass flow, h enthalpy, $\dot{Q}$ the rate of heat transfer and $\dot{W}$ the power developed or provided.

$$\sum_{in} \dot{m} = \sum_{out} \dot{m} \tag{2}$$

The modeling cycle is divided into two parts, one is the generation of the connection topology of the devices, which was performed using graph theory, and the second is the generation of algebraic equations based on the model of each device and their interconnections.

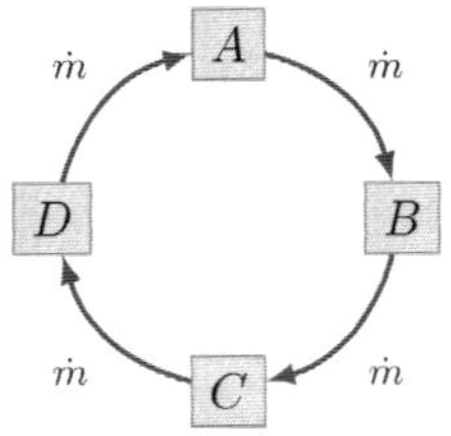

Figure 1. Sequential simulation.

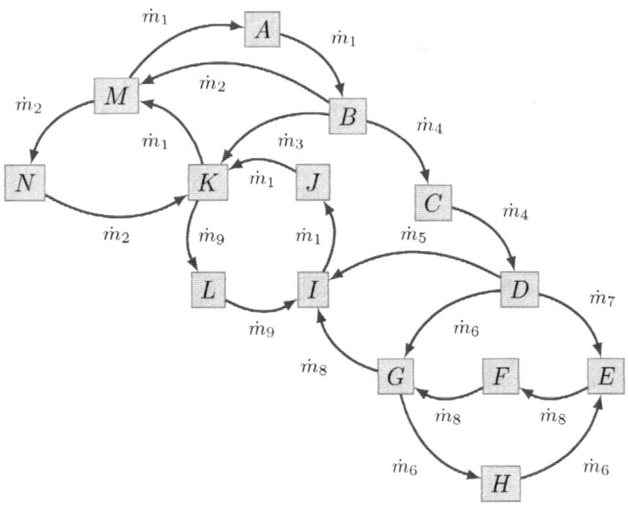

Figure 2. Simultaneous simulation.

Simulation methods cyclic processes can be classified into two categories [11]; sequential simulation or simultaneous simulation. In the sequential simulation as shown in Figure 1, the output of a component is input for the next component, there is no need to iterate and the same value of mass flow circulated throughout the cycle. In a simultaneous simulation as shown in Figure 2, there are many interconnections between the components and the mass flow through each can be different, involving a solution of a system of equations and in the case of not linear it is necessary to perform iterative process.

The device connections was performed by graph theory, which facilitates the sequential calculation devices. For example for the cycle shown in Figure 3, the equivalent graph is shown in Figure 4.

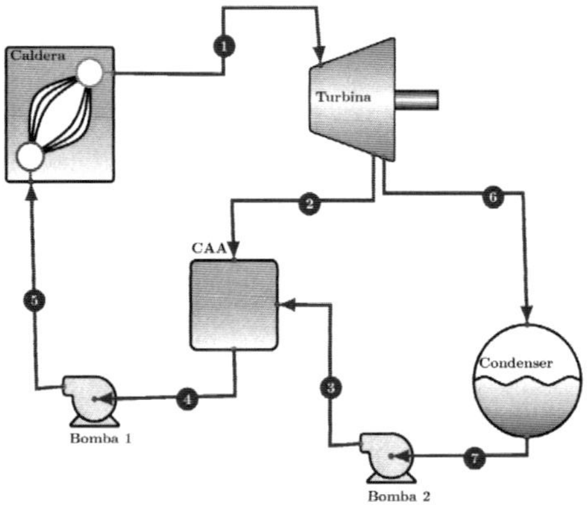

Figure 3. Diagram example of a plant.

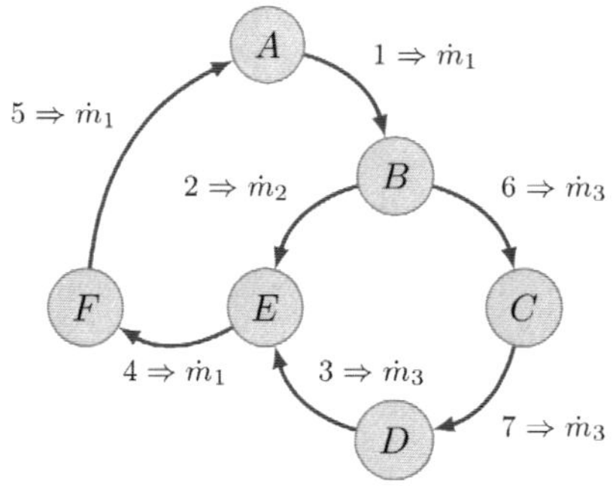

Figure 4. Graph device connections.

Where A corresponding to the boiler, B to the turbine C to the condenser, D to the pump 2, E to the CAA and F to the pump 1. This is a directed graph and it has an associated adjoint matrix as shown in Figure 5, where for a matrix element $a_{i,j}$ if the values is 0, there is no connection between the devices i and j, otherwise if the value of the item is different from 0 the device i connects to the device j with a mass flow of $a_{i,j}$.

$$
\begin{array}{c} \\ \\ A \rightarrow \\ B \rightarrow \\ C \rightarrow \\ D \rightarrow \\ E \rightarrow \\ F \rightarrow \end{array}
\begin{array}{c}
\begin{array}{cccccc} \mathrm{A} & \mathrm{B} & \mathrm{C} & \mathrm{D} & \mathrm{E} & \mathrm{F} \\ \downarrow & \downarrow & \downarrow & \downarrow & \downarrow & \downarrow \end{array} \\
\begin{bmatrix}
0 & \dot{m}_1 & 0 & 0 & 0 & 0 \\
0 & 0 & \dot{m}_3 & 0 & \dot{m}_2 & 0 \\
0 & 0 & 0 & \dot{m}_3 & 0 & 0 \\
0 & 0 & 0 & 0 & \dot{m}_3 & 0 \\
0 & 0 & 0 & 0 & 0 & \dot{m}_1 \\
\dot{m}_1 & 0 & 0 & 0 & 0 & 0
\end{bmatrix}
\end{array}
$$

Figure 5. Adjoint matrix.

3. Simulation

Simulating a cycle consists of four sequential processes. First in the GUI interface is designed cycle, then the GUI generates an XML file, which is used as input to the main application and finally the report is generated in PDF format.

The application responsible for the simulation takes as input the document *XML* created and generates a report with the results in *pdf* format, through the LaTeX tool. This application was made in `C++` and it has two major classes, the class `Planta` and the class `Dispositivos`. The class diagram `Planta` is shown in Figure 6, it shows how this class makes use of the class `CiclosXML` designed to interact with files *XML* and save this information to a member of the class `Planta`. One member of the class `Planta` is a list of objects of class `Dispositivos` shown

IOP Conf. Series: Materials Science and Engineering **138** (2016) 012017 doi:10.1088/1757-899X/138/1/012017

in Figure 7 which is an abstract class that provides the basis for other classes that implement the various methods of each of the specific devices.

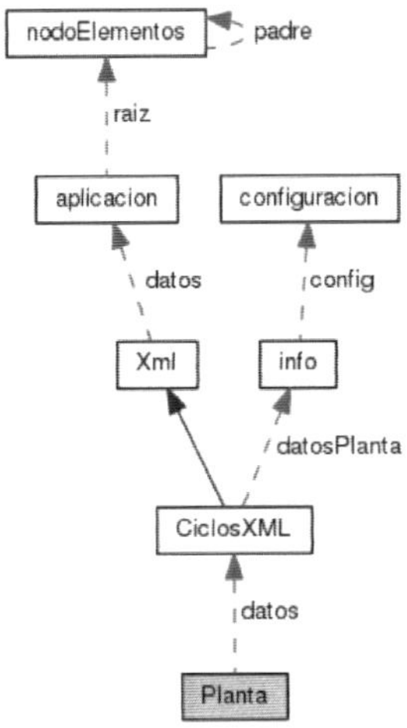

Figure 6. Diagram for the class Planta

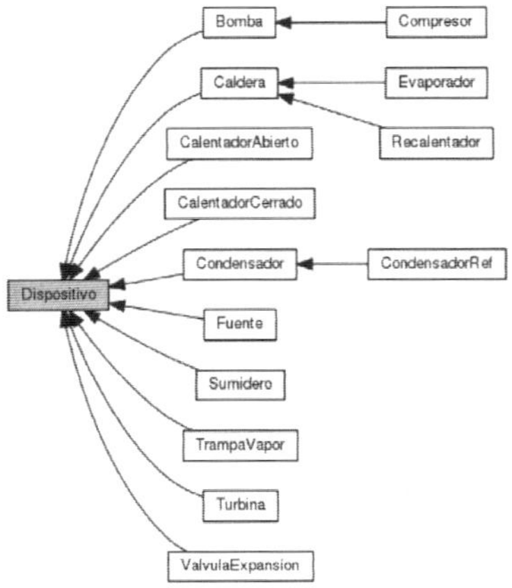

Figure 7. Diagram for the class Dispositivos

4. Results and discussion

The developed model was validated with examples found in the literature, one for a refrigeration system and another for a Rankine cycle. The refrigeration cycle used for validation is shown in Figure 8, with the following inputs: fluid R134A, $P_1 = 140\,kPa$, $x_1 = 1$, $P_2 = 800\,kPa$, $P_3 = 800\,kPa$, $x_3 = 0$ and $P_4 = 140\,kPa$.

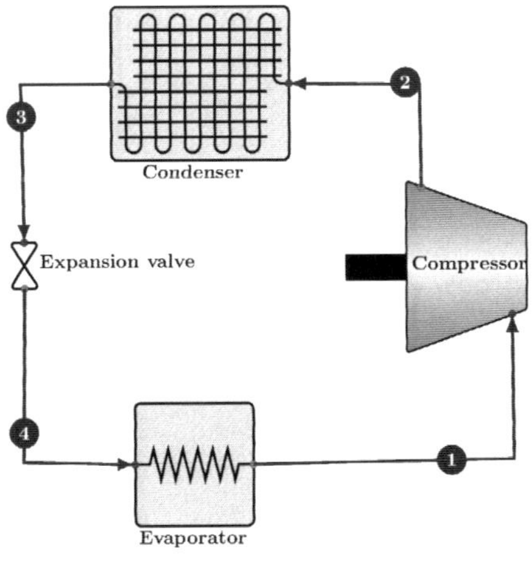

Figure 8. Diagram of the refrigeration cycle

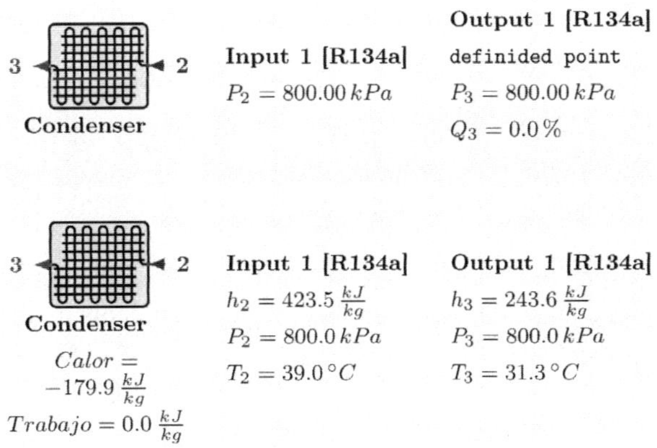

Figure 9. Data input and output for the condenser

The coefficient of performance obtained in the calculations was:

$$\beta_{Refrigerator} = 3.968$$

Table 1. Properties of each point

State	Fluid	Flow[frac]	Pressure [kPa]	Temperature [K]	Enthalpy [kJ/kg]
1	R134a	1.000	140.000	254.390	387.320
2	R134a	1.000	800.000	312.127	423.532
3	R134a	1.000	800.000	304.477	243.645
4	R134a	1.000	140.000	254.390	243.645

The summary of the properties of each point of the refrigeration cycle is shown in Table 1, and the Figure 9 shows example data of calculus for the refrigerator.

The validation for a thermal plant is shown in Figure 10, which is a Ideal Rankine cycle with the following input data: fluid water, $P_1 = 7000\,kPa$, $T_1 = 600°C$, $P_2 = 3000\,kPa$, $P_3 = 1800\,kPa$, $P_4 = 800\,kPa$, $T_5 = 500°C$, $P_6 = 100\,kPa$ and $P_7 = 5\,kPa$.

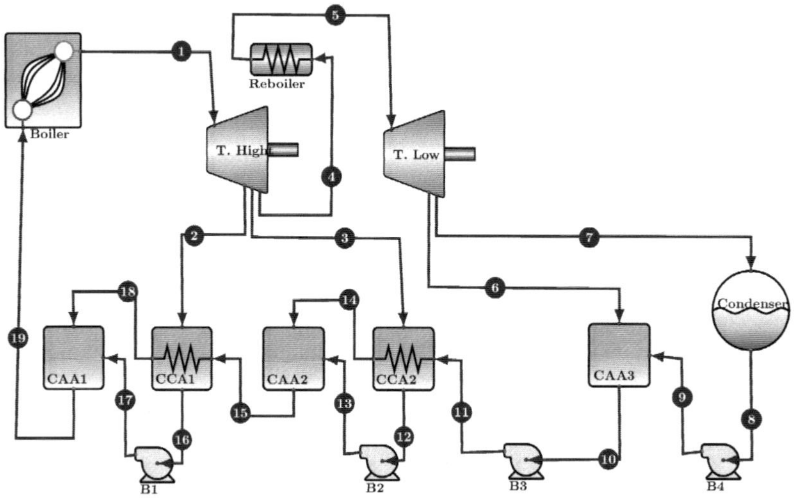

Figure 10. General diagram of the plant

The thermal efficiency of the power cycle was obtained:

$$\eta_{thermal} = 47.052\,\%$$

The summary of the properties on each of the points is shown in Table 2. With these two examples is demonstrated the correct operation of the software tool, which simulated a refrigeration cycle and complex Rankine cycle. In the Rankine cycle, the program calculated the seven different streams found in the cycle as shown in the "Flow" column of Table 2.

5. Conclusions

This computational tool facilitates the configuration of different topologies of power cycles and refrigeration cycles because it performs a series of calculations a quick and easy way. Perform these same calculations manually involve a lot of time in the calculation of thermodynamic properties and in solving a system of equations. The developed program has the ability to simulate different power cycles with different working fluids, for this reason it is possible to simulate organic cycle Rankine power and cooling cycles and it was demonstrated that the program can solve power cycles have some complexity as shown in the results.

Table 2. Properties of each point of the thermal plant

State	Fluid	Flow[frac]	Pressure [kPa]	Temperature [K]	Enthalpy [kJ/kg]
1	Water	1.000	7000.000	873.150	3650.594
2	Water	0.050	3000.000	724.872	3348.719
3	Water	0.158	1800.000	645.915	3191.964
4	Water	0.792	800.000	535.685	2977.341
5	Water	0.792	800.000	773.150	3481.253
6	Water	0.080	100.000	481.278	2891.511
7	Water	0.712	5.000	306.024	2400.193
8	Water	0.712	5.000	306.024	137.768
9	Water	0.712	100.000	306.026	137.842
10	Water	0.792	100.000	372.756	417.500
11	Water	0.792	7000.000	373.230	424.692
12	Water	0.158	1800.000	480.262	884.482
13	Water	0.158	7000.000	481.176	890.518
14	Water	0.792	7000.000	480.262	886.399
15	Water	0.950	7000.000	480.414	887.086
16	Water	0.050	3000.000	507.003	1008.353
17	Water	0.050	7000.000	507.886	1013.143
18	Water	0.950	7000.000	507.003	1009.008
19	Water	1.000	7000.000	507.047	1009.213

References

[1] F Tchanche B, Pétrissans M and Papadakis G 2014 *Renewable and Sustainable Energy Reviews* **39** 1185–1199 ISSN 1364-0321

[2] Jeon S G and Son G 2005 *Systems Modeling and Simulation: Theory and Applications* (Springer) pp 222–229

[3] Shirakawa M 2006 *Proceedings of the Institution of Mechanical Engineers, Part A: Journal of Power and Energy* **220** 569–579 ISSN 0957-6509, 2041-2967

[4] Fang H, Chen L, Dlakavu N and Shen Z 2008 *IEEE Transactions on Energy Conversion* **23** 834–841 ISSN 0885-8969

[5] DiGenova K K J 2011 *Design of organic Rankine cycles for conversion of waste heat in a polygeneration plant* Thesis Massachusetts Institute of Technology thesis (S.M.)–Massachusetts Institute of Technology, Dept. of Mechanical Engineering, 2011.

[6] Kim D W, Youn C, Cho B H and Son G 2005 *International Journal of Control Automation and Systems* **3** 493

[7] Akolekara H D, Srinivasan P and Challac J S 2014 *Int. J. of Thermal & Environmental Engineering* **8** 55–61

[8] Bell I H, Wronski J, Quoilin S and Lemort V 2014 *Industrial & Engineering Chemistry Research* **53** 2498–2508 ISSN 0888-5885, 1520-5045

[9] Lemmon EW, Huber ML and McLinden MO 2013 NIST Standard Reference Database 23: Reference Fluid Thermodynamic and Transport Properties-REFPROP

[10] Borgnakke C and Sonntag R E 2012 *Fundamentals of Thermodynamics* 8th ed (Hoboken, NJ: Wiley) ISBN 978-1-118-13199-2

[11] Panosso G C and Schneider P S 2003 (Sao Paulo, Brasil)

Sequential reductive and oxidative conditions used to biodegradation of organochlorine pesticides by native bacteria.

M Kopytko[1]*, S N Correa-Torres[1] and A Plata[2]

[1] Environmental Engineering Faculty, Universidad Pontificia Bolivariana, Bucaramanga. Piedecuesta, Km 7.0. Off. K-107. Bucaramanga, Colombia.

[2] Biotechnological Caribbean Center, National Learning Service, Regional Cesar, La Paz, Km 7.0, Valledupar, Colombia.

E-mail: maria.kopytko@upb.edu.co; sandra.correa@upb.edu.co; amplata4@misena.edu.co

Abstract. Despite restrictions and bans on the use of many organochlorine pesticides in the 1970s and 1980s, they continue to persist in the environment today. This is the case of Agustin Codazzi, Cesar where the organochlorine pesticides were buried without control in the soil in 1999, after being banned their use. Nowadays is necessary to find the best method, which allows remediation of this soils. Reductive dechlorination is the first and limiting step in the metabolism of many organochlorine pesticides by anaerobic bacteria. In this study the reductive conditions were enhanced by addition of biogas as an auxiliary electron donors.The soil sample was taken from the zone at Agustín Codazzi, Cesar, and their characteristics correspond to a loam soil with low nutrient and slight compaction. The experimental tests were performed by varying the exposure time of a reducer to oxidative environment. Reductive conditions were enhanced by methane from biogas and oxidative environment was generated by air blown to stimulate a metabolic process of the soil native bacteria. Removals between 70 and 78.9% of compounds such as 4,4'-DDT, 4, 4'-DDD, 4,4'-DDE, Endrin and Trans-Chlordane, detected by gas chromatography analysis, were achieved under reductive/oxidative conditions during 120 days. Furthermore, bacterial strains capable of degrading organochlorine pesticides were selected from the native bacteria, and identified by the purified and identified based on its morphological characteristics and 16S rDNA sequencing.

1. Introduction

The introduction of agrochemical compounds for the elimination of crop pests in the past decades has generated a group of pollutants tending to persist in the environment. Organochlorine pesticides such as DDT, Toxaphene, DDT and Endrin, among others, have caused world disturbance since it was learned about how harmful they are to human health and the environment. Have been well documented that decades after application to soil, organochlorine pesticides keep on being detected in the aquatic and terrestrial species including humans [1-3]. In 1997 the Board of Directors of the United Nations Environment Program (UNEP) started negotiations with the Intergovernmental Forum on Chemical Safety (IFCS) aiming at reducing the risks entailed by these compounds, through the creation of the first list of twelve Persistent Organic Pollutants (POP) [4]. In Colombia there are still some sequels of the inappropriate use and store of organochlorine compounds. During the 1960's and

[1]* To whom any correspondence should be addressed

IOP Conf. Series: Materials Science and Engineering **138** (2016) 012018 doi:10.1088/1757-899X/138/1/012018

the 1980's the Department of Cesar was consolidated as the main cotton producer and exporter in the country [5]. From 1990 to 1994 Corporación Autónoma Regional del Cesar (Corpocesar), Geosísmica y Ambiente, and Cesar's Secretariat of Agriculture carried out soil surveys which revealed that many of the agrochemicals (mainly organochlorine pesticides) were buried, with no control in much of the land adjacent to the growing areas, like the recently created Instituto Técnico Agropecuario Antonio Galo Lafaurie, currently abandoned [6]. Most cases of cancer and degenerative diseases occurred since then have been attributed to such fact, as well as the increased mortality rate. Even so, up to the present there has been no solution to remedy the impact caused by this pesticide waste deposits. In literature has been recognized the need to carry out sequential anaerobic - aerobic treatment of the organochlorine compounds [7-8]. The halogenated atoms of the molecule of these compounds increase the carbon oxidation states and the aerobic processes are less favorable energetically for highly halogenated compounds. For that reason, the anaerobic degradation results more convenient in the first phase since the more halogens the molecule has, the easier it is to produce a reductive dehalogenation. Once the pesticides are dechlorinated, they produce metabolites more susceptible to further aerobic degradation. Considering the anaerobic phase as the most efficient, stimulation in reducing conditions has been sought for the transfer of the electrons required in the dechlorination process [9]. Usually, organochlorine pesticides inhibit the growth of microorganisms, however, long permanence of these compounds in the soil contributes to microorganisms' adaptation to this sort of substances and their usage as the primary source of carbon during biodegradation process [10-12].
The purpose of this research was to verify the viability of improving biodegradation in the first phase in which native bacteria were supported with the methane, produced by biogas and injected to the soil treated, as an auxiliary electron donor. Later, in a sequential way, air was applied to the same sample of polluted soil in order to continue the sequential anaerobic-aerobic process of degradation of these pesticides. At the end of each phase it has been carried out a chromatographic analysis of the remnant concentration of organochlorine compounds in the treated soil. In the same way native bacteria were isolated from soil samples from an illegal deposit of organochlorine pesticides in the locality of Agustin Codazzi, Cesar.

2. Experimental Setup

The soil compound sampling was carried out in Agustín Codazzi, Cesar (See Figure 1), in the earliest facilities of Instituto Técnico Agropecuario Antonio Galo Lafaurie, a zone which registered the existence of a great variety of pesticides concentrations, according to reports of Corporación Autónoma Regional del Cesar (Corpocesar).

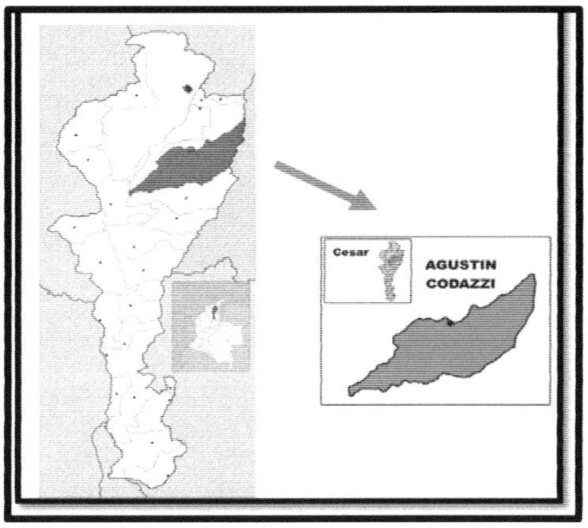

Figure 1. Sketch map of sampling site, Agustín Codazzi, Cesar, Colombia [13]

Agustín Codazzi is finds located in the north part of the Department of Cesar where the annual average of temperature is 29.5 °C (85.1 °F), with maximum and minimum range of 24 and 35 °C (75 and 95 °F) [13].

2.1. Sampling

For the sampling, the specifications of Corpoica were applied [14] where first was delimited the zone in two quadrants of 50x50 meters; after that zig-zag points were located every 12 meters. In each point the samples were collected at 40 cm below ground level, in total 80 units of approximately 3 kg each. Later, the compound samples were prepared through crushing, screening and homogenization.

2.2. Initial conditions of soil.

In order to get the initial characteristics of soils, physical properties such as texture, humidity, density, total porosity and field capacity were determined. The chemical tests were pH, conductivity and cation-exchange capacity (CEC), Organic Material (OM) and Nutrients (N and P). In all analyzes were applied standardized methods and/or methods proposed by the regulations of American Society of Agronomy (ASA), American Society for Tests and Materials (ASTM) and Colombian Institute of Technical Standards and Certification (ICONTEC), among others.

Later, the native bacterial genres were isolated and identified, adapted to the conditions of the polluted soil, by means of macro description and microscopic description of isolated strains, the application of the analytical profile index for non-enteric bacteria (API 20 NE), as well as the molecular identification by PCR of the region 1465 pb of the 16S ribosomal gene.

The Matrix Solid Phase Dispersion (MSPD) was used for the detection and identification of organochlorine compounds, by applying an AT chromatographic ® 6890 Series Plus adapted to a mass detector, which was operated in SIM mode (Selected Ion Monitoring). A DB-5MS column (5%-phenil-poli (methyl siloxane), 60 m x 0.25 μm), and the injection was carried out in Splitless mode with a 2μL volume. For the detection of these compounds was used as reference material the certified mixture of organochlorine pesticides Part N° OCP508-1JM (ChemService, West Chester, PA 19381 PO BOX 599, US).

2.3. System dimensioning

8.5 kg of soil sampling were taken in a plastic container covered with Kraft paper externally in order to avoid the passage of light. Inside the recipient were put three PVC pipes of half an inch (½") perforated in flute style. A fitting was adapted to each end; and such end was connected to a hose protruding to the surface. This hose was first connected to an anaerobic digester in order to pump biogas; and then it was connected to an air pumping system to comply with the reducing and oxidative conditions, respectively. Three reactors were subjected to 40 days of Biogas plus 80 days of air. They were called RT1. The other three were subjected to 80 days of Biogas and 40 days of air, and were named RT2. Additionally, the natural attenuation control (NA) was taken into account.

2.4. Biogas quantification and dimensioning of the treatment system

Initially a test of specific methanogenic activity was carried out, by using fresh cow manure taken from the gardening area at Universidad Pontificia Bolivariana. A solution was prepared with a concentration of 2.5 g total volatile solid/L (TVS/L) which is considered to be the optimal concentration to reach the maximum methane generation capacity [15]. Once the biogas amount was quantified as well as the time necessary to renew its supply, the reactors assembly was carried out. On average, along with the six inlets, approximately 120 cm^3 of methane were injected to each reactor per day. Simultaneously there was a daily watering keeping between 50% and 80% of the field capacity, humidity required for the activity of microorganisms. Once the reducing phase was finished, the sequential oxidative phase was applied in the same conditions by inducing air to the reactors instead of biogas. Each air inlet provided a flow of 178,4 m^3 per day, which was equivalent to 1070 m^3 of air per recipient every day.

3. Results and discussion

3.1. Determination of initial soil conditions

The soil used for the study was loamy (average ± standard deviation) 45,76% ± 3,05% sand, 41, 28% ± 3,28% silt, and 12,96& ± 1,65% clay. The apparent density was 1.604 g/cm3 ± 0,02 and 23,45% ± 4,2% porosity. The soil had low humidity equal to 2,43% ± 0,2%. The low concentration of organic material corresponding to 1,08%, Nitrogen of 1025,55 mg /kg of soil, and Phosphorus 1,33 mg/kg of soil. The cation-exchange capacity (CEC) was 17,6 ± 0,52 meq-g/100g soil and the pH was 6,68 ± 0,03. Finally, the conductivity was 204,6 µs/cm ± 6,6 with high chloride concentration of 4008,3 mg/l. The soil showed five (5) organochlorine compounds, whose concentration is reported in table 1.

Table 1. Initial concentration and chromatogram of organochlorine compounds in the soil of interest.

Organochlorine compounds	Retention time (min)	Concentration (mg/kg soil)
trans-Chlordane	40,30	0.1
4.4'-DDE	42,12	1.1
Endrin	43,44	1.9
4.4'-DDD	44,14	2.4
4.4'-DDT	45,74	9.1

Relating high concentrations of chloride present in the soil studied and high concentrations of metabolites of 4, 4'-DDT (4,4'-DDD and 4,4'-DDE) present, it is considered a progress in the ongoing biodegradation process, carried out by native bacteria adapted to the conditions of the place. In the same way, the great amount of DDT and Endrin found in the soil sample, where these pesticides were buried more than 15 years ago, indicates the need to look for an efficient method which can accelerate its natural biodegradation process.

In the initial microbiological identification of native bacteria present in the polluted soil were identified 37 strains of the following genres: *Pseudomonas, Aeromonas, Burkholderia, Bacillus and Enterobacter* of which one of identified species (2F) are illustrated in Figure 2. These bacterial species (*Pseudomaonas*) are reported in literature as strains with a high potential to degrade organochlorine compounds [16-17].

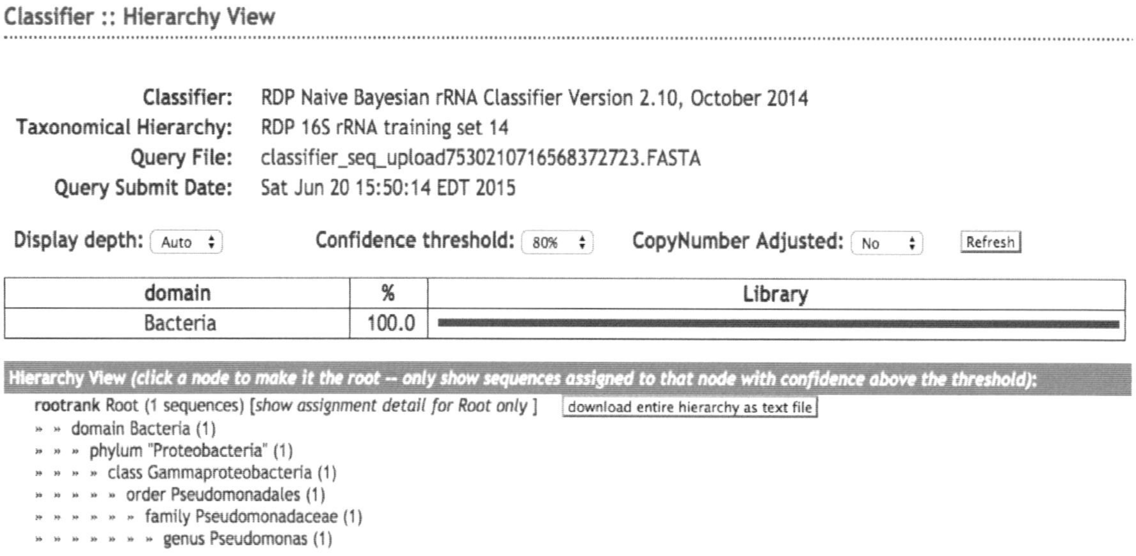

Figure 2. Result of the comparison of the sequence problem of against the database RDP

The search in the database Greengenes indicates that the sequence problem is very similar to sequences of *Pseudomonas sp.*and *Pseudomonas putida, Pseudomonas fluorescens* and *Pseudomonas koreensis*.

Nevertheless the sequence problem assembled of 1457 pb against the database treated of the NCBI, Genomic Reference Sequences, RefSeq, indicates that it has 99,9 % of identity and 100 % of his length, with sequences of the gene ribosomal 16s, belonging to *Pseudomonas sp.* and *Pseudomonas fluorescens.*(Figure 3)

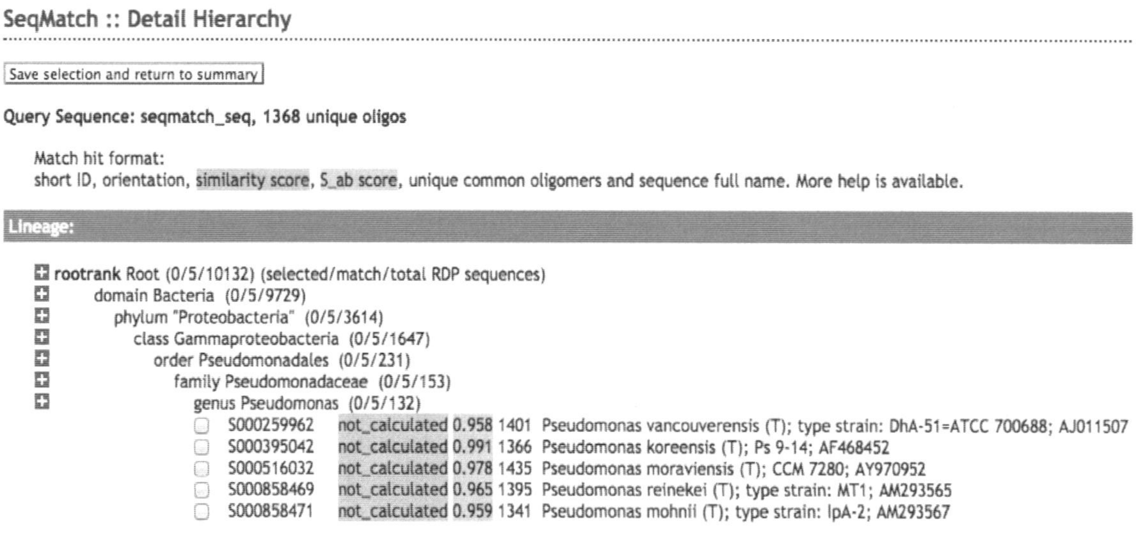

Figure 3. Sequence problem assembled of 1457 pb against the database treated of the NCBI

The tree of distances (Figure 3) constructed from the 30 most nearby, available sequences of cultivable microorganisms in the database nr/nt of NCBI, it shows that the sequence problem is located by

Pseudomonas sequences of the genre, with a support of branches of 95 %. but the tree does not allow to classify up to species.

3.2 Treatment of the soil sample in reducing and sequential oxidative conditions
Figure 4 summarizes the results obtained after bioremediation in reducing sequential oxidative conditions, with the application of biogas and air in the respective phases and native bacteria present in the polluted soil.

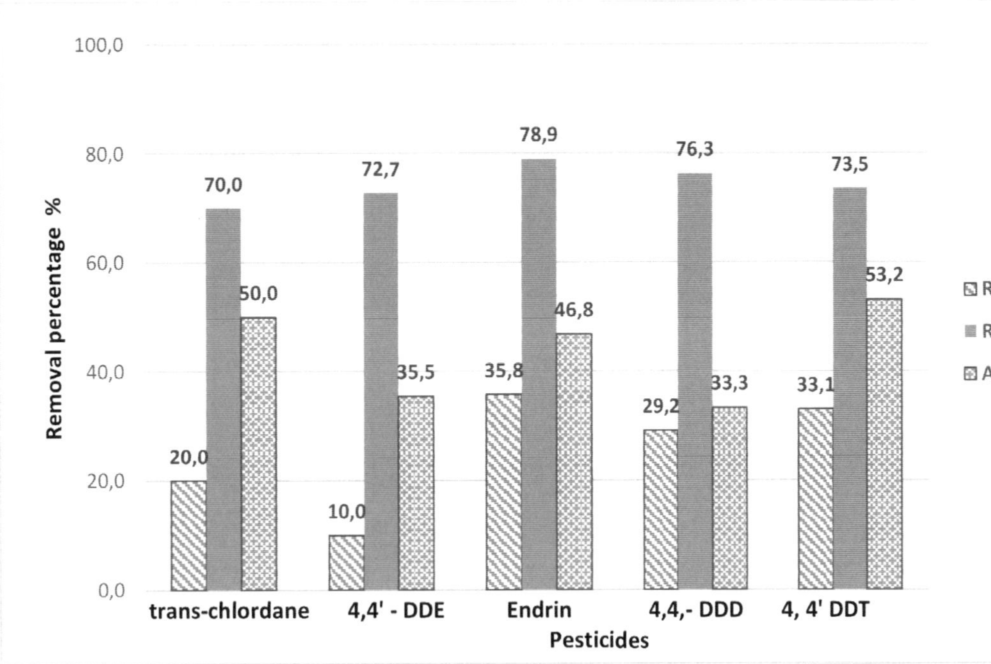

Figure 4. Removal percentage of organochlorine pesticides from soil used in the study.

The bioremediation process lasted 120 days, distinguishing high removal of organochlorines in RT2 reactors, where the time of exposure to reducing conditions was 80 days, followed by 40 days of oxidative phase. The best result was shown by Endrin with a removal percentage of 78,9% followed by 76,3% of 4,4'DDD removal, and 73,5% of DDT, removal in the same reactors. Considering that trans-chlordane and Endrin belong to a group of chlorinated cyclodienes and DDT, and its metabolites (DDD and DDE) are chlorinated aromatics, there is a difference between the stability and the shape of the structures they represent, which makes a difference in the bioavailability of these compounds and their consequent biodegradation. The results obtained indicated an unfavorable condition of short term applied during the reducing phase (40 days) developed in the RT1. For these reasons, it is corroborated that this stage is crucial for the dehalogenation of organochlorine compounds.

4. Conclusions

The results obtained from the soil physicochemical characterization provide an indication of the deterioration of its properties. Although the soil texture is considered to be optimal for root growth, the apparent density value (1.604 g/cm^3) and total porosity (23.45%) evidence a degree of light to medium compaction, caused by the retention of pollutant in the soil pores. The Nitrogen and Phosphorus concentrations (1025.55 and 1.33 mg/Kg of soil, respectively) as well as low levels of cation exchange (17.6 meq- g/100 g Soil) and organic material content (1.08%) reveal poor soil that is unbalanced in essential nutrients. The soil analyzed shows a selected group of bacteria adapted to conditions of pollution by organochlorines, of which exist bibliographical references as for degradation potential of these compounds such *as P. Putida, P. Fluorescens, B. Cepacia, E. Coli, Bacillus sp, A. Xylosoxidans, B. Cepacia,* among others. The reducing phase is fundamental to the degradation of the organochlorine pesticides present in the soil analyzed, which is why greater percentages of removal of organochlorines were obtained in the RT2 reactors which remained more time in reducing conditions (80 days) and less time in oxidative conditions (40 days). The best removal result was shown by Endrin (78,9%), followed by 4,4'DDD (76,3%) in RT2. The degradation of organochlorine compounds in soil, under natural conditions, is a slow process. Therefore, the stimulation of microbiota turns out to be favorable through the generation of reducing and oxidative sequential conditions to accelerate the biodegradation process.

Acknowledgments
The authors acknowledge the financial support of this work through Project DGI 053-0614-2400 (Universidad Pontificia Bolivariana- seccional Bucaramanga)

References

[1] Miersma N, Christopher, A, Pepper, B, Anderson, T, 2003 Organochlorine pesticide in elementary school yards along the Texas–Me´xico border. Environmental Pollution 28, 65–71.
[2] Borga K, Gabrielsen, G, Skaare, J, 2001 Biomagnification of organochlorines along a Barents Sea food chain. Environmental Pollution 113, 187-198.
[3] Chopra, A, Sharma, M, Chamoli, S, 2011 Bioacumulation of organochlorine pesticides in aquatic system-an overview. EnvironMonitAssess 171:905-916.
[4] PNUMA 2005 Eliminando los COP del mundo: Guía del convenio de Estocolmo sobre contaminantes orgánicos persistentes.
[5] Barrios M, 2013 En Codazzi solo quedan ruinas de la bonanza algodonera. *El Heraldo* pág. 1.
[6] CORPOCESAR & Geosísmica y Ambiente 1997 Estudio de diagnóstico y evaluación del enterramiento de pesticidas en el municipio de Agustín Codazzi y procedimiento de descontaminación y disposición final de los mismos. Valledupar, Cesar, Colombia.
[7] Foght J, April T, Biggar K, Aislabie J, 2001 Bioremediation of DDT-contaminated soils: a review Bioremed. J. 5 pp. 225–246
[8] Langenhoff A,Pijls CH, Alphenaar A, Zwiep G, Rijnaarts H, 2002 Intrinsic and stimulated in situ biodegradation of hexachlorocyclohexane (HCH) Water Air Soil Poll. Focus 2. pp. 171–181
[9] Baczynski T, Grotenhuisb T, Knipscheerb P,2003 The dechlorination of cyclodiene pesticides by methanogenic granular sludge Chemosphere 55 pp. 653–659
[10] Fang H, Dong B, Yan h, Tang F, Yu Y, 2010 Characterization of a Bacterial Strain Capable of Degrading DDT Congeneres and its Use in Bioremediation of Contaminated Soil. Journal of Hazardous Material, Vol. 184, pp. 1-3.
[11] Betancur-Corredor B, Pino N, Peñuela G, Cardona-Gallo S, 2013 Bioremediation of a pesticide polluted soil: case DDT. Revista Gestión y Ambiente,Vol. 16 (3) pp. 119-135.
[12] Li XM, Lin Z, Luo CL, Bai J, Sun YT, Li YT, 2015 Enhanced microbial degradation of pentachlorophenol from soil in the presence of earthworms: Evidence of functional bacteria using DNA-stable isotope probing. Soil biology &Biochemistry 81, pp.168-177
[13] CORPOCESAR, & Geosísmica y Ambiente. 1997 Estudio de diagnóstico y evaluación del enterramiento de pesticidas en el municipio de Agustín Codazzi y procedimiento de descontaminación y disposición final de los mismos. Valledupar, Cesar, Colombia.
[14] Corporación Colombiana de Innvestigación Agropecuaria CORPOICA. 2014 Guía para la toma de muestras de suelo.

[15] Torres P, Pérez A, Cajigas A, Jurado C, Ortiz N, 2007 Selección de inóculos para el tratamiento anaerobio de aguas residuales del proceso de extracción de almidón de yuca.Ingeniería de Recursos Naturales del Ambiente, 6 pp. 105-111.

[16] Nawab A, Aleem A and Malik A, 2003 Determination of organochlorine pesticides in agricultural soil with special reference to γ-hch degradation by *Pseudomonas* strains. Science direct, 41-46.

[17] French WT, Brown LR, Downer DN, Fredrickson HL, Teeter CL, 2002 Effects of n-hexadecane and pm-100 clay on trichloroethylene degradation by *Burkholderia cepacia.* Science direct 89-102.

Automatic humidification system to support the assessment of food drying processes

B D Ortiz Hernández[1], A R Carreño Olejua[1], J M Castellanos Olarte[1*].

[1] Universidad Pontificia Bolivariana Seccional Bucaramanga, Colombia.

E-mail: javier.castellanos@upb.edu.co.

Abstract. This work shows the main features of an automatic humidification system to provide drying air that match environmental conditions of different climate zones. This conditioned air is then used to assess the drying process of different agro-industrial products at the Automation and Control for Agro-industrial Processes Laboratory of the Pontifical Bolivarian University of Bucaramanga, Colombia. The automatic system allows creating and improving control strategies to supply drying air under specified conditions of temperature and humidity. The development of automatic routines to control and acquire real time data was made possible by the use of robust control systems and suitable instrumentation. The signals are read and directed to a controller memory where they are scaled and transferred to a memory unit. Using the IP address is possible to access data to perform supervision tasks. One important characteristic of this automatic system is the Dynamic Data Exchange Server (DDE) to allow direct communication between the control unit and the computer used to build experimental curves.

1. Introduction

Food drying is an industrial process used to reduce water activity in products allowing its conservation, storage stability and transportation [1]. However, the development of new drying techniques or the evaluation of new dried products requires a very careful approach. All processes involved even in simpler dryers are highly non-linear and its scale-up is generally very difficult, requiring experimentation at laboratory and pilot scales coupled with the experience of the researcher [2]. Hence, a research approach involves basic models that are validated using experimental outcomes. Most of models require an understanding of the typical product behavior during drying, and most of this information can be drawn from the so-called drying curves that presents the mass decrement versus time at stable drying conditions. Every food has a representative drying curve related to drying air velocity, temperature and pressure that reflects how removal of water is affected during the process; using the slope of these curves, is possible to obtain the drying rate curve, which is in turn very important to make assumptions for the construction of mechanics of dried materials [3,4]. Foods in particular presents, after a period of constant rate drying, a period characterized by a deceleration of water removal. The rate-controlling factors in this period are complex, depending upon diffusion through the food, and upon the changing energy-binding pattern of the water molecules. There is little theoretical information available for drying of foods in this particular region and experimental drying curves are the only adequate approach.

*To whom any correspondence should be addressed.

The drying curves are based on the mass reduction with time when the product is dehydrated in a laboratory dryer. Most of the experimental devices used are composed by a propeller fan, an electric heater, a scale and a humidifier. The temperature and the moisture content of the drying air is keep stable through heating and a supply of fresh air or steam, or using saturated air.

One of the most important parameters in drying is humidity, but unlike temperature and velocity of air, it is quite difficult to measure it and control it. In earlier experiments the air used had a low humidity or was taken from the environment. This is a very simple way to set up experiments, but it has a poor reproducibility because the humidity depends on ambient conditions [5]. Nonetheless, it is possible to use air with known humidity. From one hand, air is equilibrated with a saturated salt solution at a constant temperature, providing a known humidity air [6]. Its drawbacks are the long time required to reach equilibrium and the difficulty to produce large amounts of conditioned air. Another procedure is known as the two-flow method [7]. In this instance, a stream of gas is divided into two parts, one is saturated with water at a certain temperature and the other is a dry gas; both streams are then mixed producing the desired conditioned air. Its most important drawback is the flow rate measurement and requires an efficient saturation process along with good temperature control of the air coming out the saturator. Similarly, in the two temperature method a stream of air is saturated with air at a given temperature and then the last is raised. Here, accuracy depends on the temperature measurement and the saturator efficiency. Finally, in the two-pressure method a stream of air at a high pressure is saturated with water at a given temperature and then reduced to a lower pressure. This method requires accurate measurement of the pressure and temperature.

As a common characteristic, all precedent methods require the measurement of at least one physical variable and their accuracy depends on how well the specific variable is measured. Besides, they can only produce limited amounts of treated air [8]. Consequently, most of the experimental units used for investigation of drying processes have a reduced size and only small samples of the product of interest can be evaluated. The experimental apparatus normally consist of a drying device in which the product sample is exposed to a controlled air flow (Temperature, humidity and speed), and both test cross section and samples zone are quite small [9,10].

2. Description of the automatic humidification system

The automatic humidification system is used to control humidity and temperature of air in an experimental facility. This facility has been developed at the Automation and Control for Agro-industrial Processes Laboratory of the Pontifical Bolivarian University, looking forward to emulate ambient air of different climatic zones and to obtain the drying curves of a wide range of agro-industrial products. Figure 1 illustrates the main elements of the test unit where the automatic humidification system was set-up. It consists of a ring of convective drying where a dryer or a drying box can be placed to evaluate different products. The unit has been instrumented with several temperature, humidity and velocity sensors for air monitoring, and strain gages to measure the weight of the product being dried.

The experimental apparatus has to produce an important quantity of conditioned air because its major purpose is to evaluate the performance of prototype dryers. In this unit, temperature and humidity of the air are controlled in real time. From one hand, it features an air conditioning unit that reduces temperature and humidity of ambient air depending on the conditions of the convective stream to be use in a particular experiment. Then, the amount the air required in the dryer is extracted by a blower and passed through the Heating Humidification Unit (H_2U), which finally delivers the air to the Dryer.

Figure 2 illustrates the main features of the H_2U. It has temperature and humidity sensors, a set of high pressure water nozzles and a set of electric heaters.

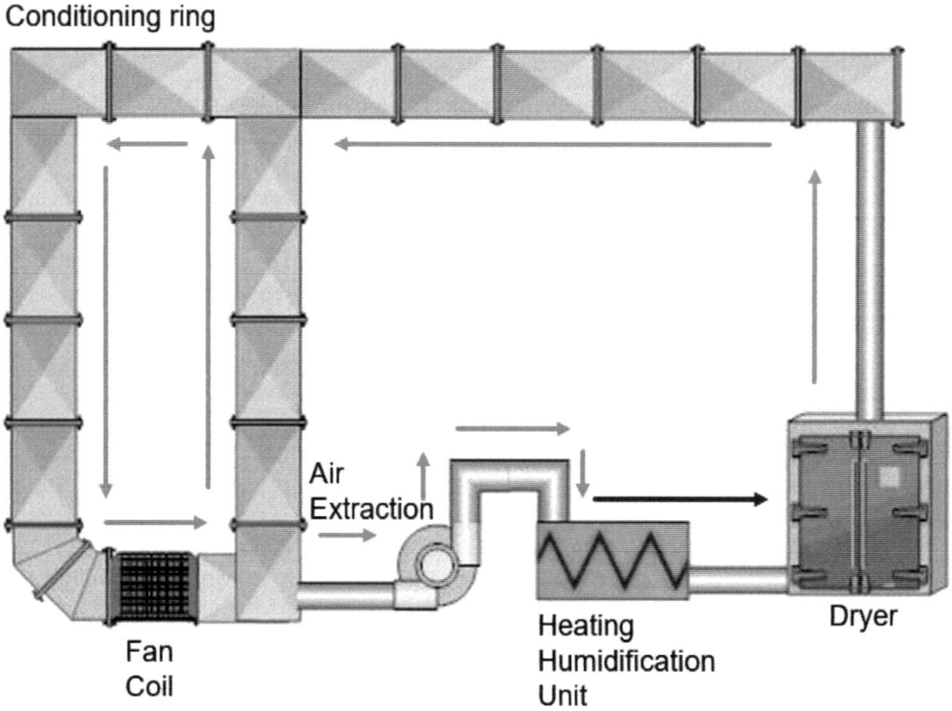

Figure 1. Experimental facility.

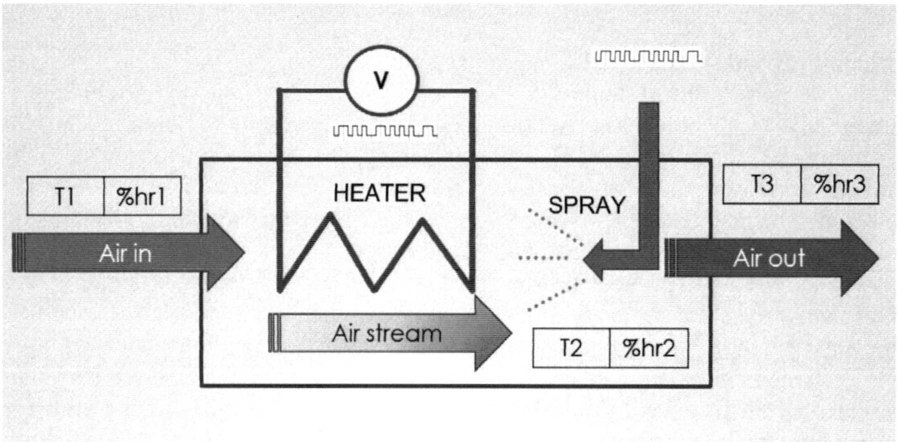

Figure 2. Heating Humidification Unit (H2U).

Sensors and actuators are linked to a control unit that is responsible for maintaining the required conditions of the air flowing to the dryer. A Programing Logic Control (PLC) device was used along with four variable frequency drives and two slave modules for in/out (IO) signals. Figure 3 shows an outline of the control network put in place and its components. The development of automatic routines to control and acquire real time data was made possible by the use of robust control systems and suitable instrumentation. The signals are read and directed to the controller memory where they are scaled and transferred to a memory unit. Using the IP address is possible to access data to perform supervision tasks.

One important characteristic of this automatic system is the Dynamic Data Exchange Server (DDE) that allows direct communication between the control unit and the computer used to build the experimental curves. Industrial devices like PLC have limited local memory, making necessary to manage and save huge amounts of data for off-line analysis. Notwithstanding that industrial systems commonly use OPC (Object Linking and Embedding for Process Control) as communication protocol, in order to link devices from different brands, it is necessity to acquire specific commercial software applications. This represents additional costs to implement supervisory control and data acquisition systems. On the contrary, DDE is a standard feature in computers running Microsoft Windows, which allows easy access to data from multiple applications. Additionally, the experimental facility has actively involved the concept of Totally Integrated Automation (TIA) to combine different automation architectures, with the derived advantages outlined in the specialized literature [11].

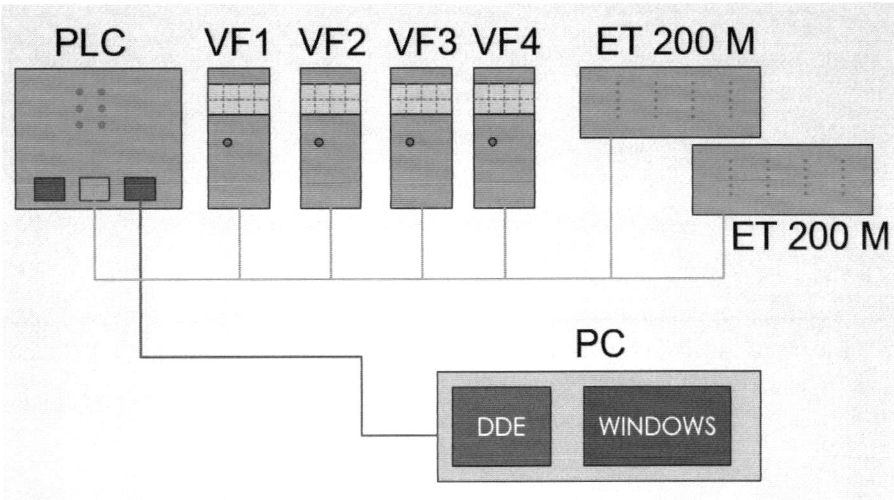

Figure 3. Control network.

3. Results

When the experimental facility is used to emulate a specific climate zone, the results show great agreement with the real air conditions. The equipment is able to produce air with the same average temperature and humidity of Colombian cities with extreme environments. Figure 4 shows a Humidity and Temperature diagram when the automatic system is controlling the variables of interest using an on/off strategy. In this instance, the temperature of the air reaches a relative humidity of 80%, while the temperature of the air leaving the H_2U unit remains at about 35°C. After this point, the unit stops humidification and the air inlet temperature is no longer controlled in the conditioning ring.

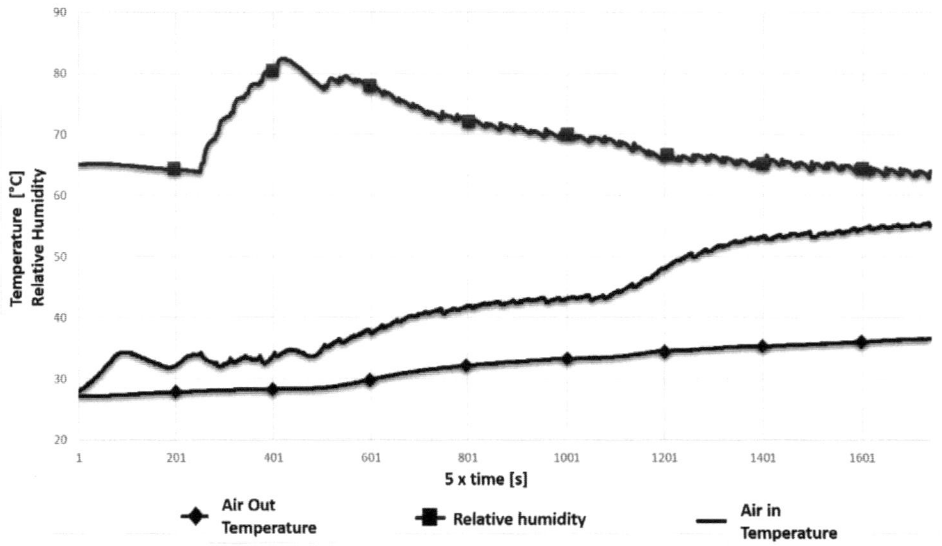

Figure 4. Humidity and Temperature of air vs time. On/off control.

Figure 5 shows another example of the system operating strategy to reach air at 50°C and of 43% relative humidity, when air temperature is high and humidity is low at the initial condition. In this instance, it was implemented a step on/off fuzzy control strategy.

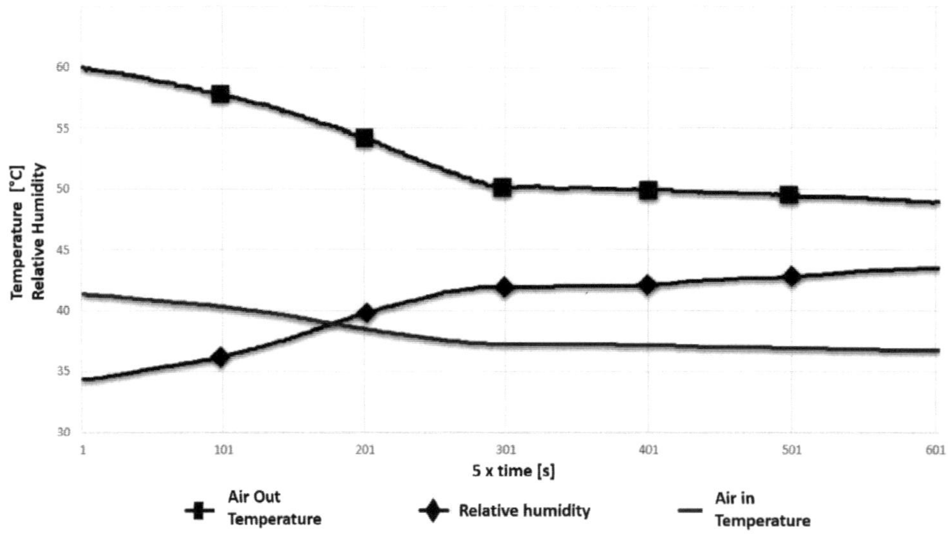

Figure 5. Humidity and Temperature diagram of air vs time. Step on/off fuzzy control.

When a prototype rotary dryer was used to build the drying curves of coffee beans and cassava, the automatic system allowed to obtain the weight of the dryer load in real time using precision strain gages. Figure 6 shows the dryer load weight variation with time. For these particular tests, environmental conditions of the air going to the dryer where set to emulate a production area of Santander in Colombia (T=28°C, RH=57%), and the results showed great agreement with other experimental studies [12,13].

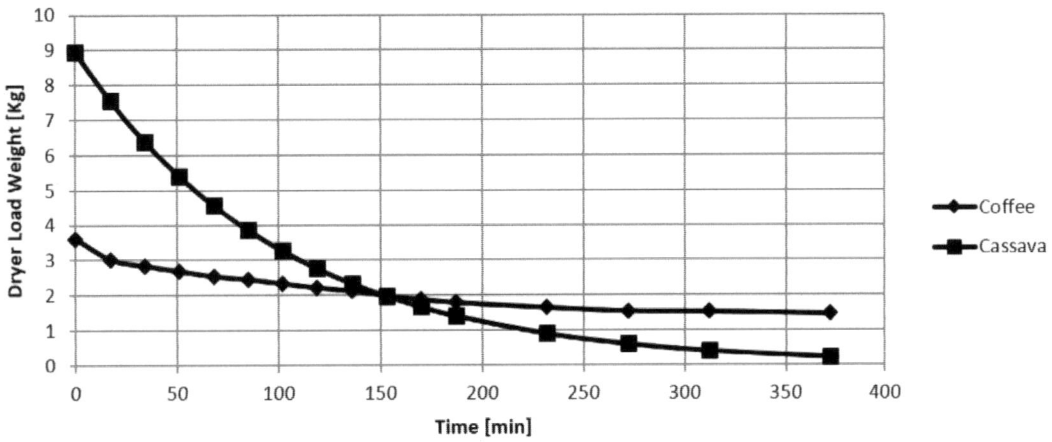

Figure 6. Variation of rotary dryer load when coupled to H_2U unit.

4. Conclusions

The automatic humidification system integrated to the experimental facility allows to configure a high capacity air conditioning unit to support drying processes, especially for the construction of drying curves of agro-industrial products. The use of a robust system to control temperature and humidity of the air for drying applications is the most important outcome. In this experimental device is possible to use different control strategies in order to obtain the particular air conditions required in the dryer under test; simple on/off control strategy combined with fuzzy control techniques showed good results and a low computational cost. Finally, the use of a DDE server to link the PLC to a computer has provided an easy and inexpensive way to process experimental data.

Acknowledgments

The authors wish to acknowledge the financial support of the Pontifical Bolivarian University of Bucaramanga and the invaluable assistance of all the people at the Automation and Control for Agro-industrial Processes Laboratory.

References

[1] Hui Y H 2008 Food drying science and technology: microbiology, chemistry, applications (Lancaster, Pa: DEStech Publ)

[2] Mujumdar A S 2015 Handbook of industrial drying (Boca Raton: CRC Press, Taylor & Francis Group)

[3] Kowalski S J 2003 Thermomechanics of Drying Processes vol 8 (Berlin, Heidelberg: Springer Berlin Heidelberg)

[4] Redman-Furey N, Normand M D and Peleg M 2013 Estimating the needed amount of desiccant, water or moistener to adjust the equilibrium water activity of dry powder mixtures *Trends Food Sci. Technol.* **29** 135–41

[5] Heinonen M 1999 A humidity generator with a test chamber system *Measurement* **25** 307–13

[6] Nguyen T A, Verboven P, Daudin J D and Nicolaï B M 2004 Measurement and modelling of water sorption isotherms of "Conference" pear flesh tissue in the high humidity range *Postharvest Biol. Technol.* **33** 229–41

[7] Di Francia G, Castaldo A, Massera E, Nasti I, Quercia L and Rea I 2005 A very sensitive porous silicon based humidity sensor Sens. *Actuators B Chem.* **111–112** 135–9

[8] Lin S X Q and Chen X D 2005 An effective laboratory air humidity generator for drying research *J. Food Eng.* **68** 125–31

[9] Jannot Y, Talla A, Nganhou J and Puiggali J-R 2004 Modeling of banana convective drying by the drying characteristic curve (DCC) method *Dry. Technol.* **22** 1949–68

[10] Tzempelikos D, Vouros A, Bardakas A, Filios A and Margaris D 2013 Design, construction and evaluation of a new laboratory convective dryer using CFD *Int. J. Mech.* **7** 425–34

[11] Samad T, McLaughlin P and Lu J 2007 System architecture for process automation: Review and trends *J. Process Control* **17** 191–201

[12] Burmester K and Eggers R 2010 Heat and mass transfer during the coffee drying process *J. Food Eng.* **99** 430–6

[13] Fernando W J N, Low H C and Ahmad A L 2011 Dependence of the effective diffusion coefficient of moisture with thickness and temperature in convective drying of sliced materials. A study on slices of banana, cassava and pumpkin *J. Food Eng.* **102** 310–6

IOP Conf. Series: Materials Science and Engineering **138** (2016) 012020 doi:10.1088/1757-899X/138/1/012020

Efficiency of modified chemical remediation techniques for soil contaminated by organochlorine pesticides

S N Correa-Torres[1], M Kopytko[1] and S Avila[1]

[1] Environmental Engineering Faculty. Universidad Pontificia Bolivariana, Bucaramanga Campus. Piedecuesta,Km 7.0. Off. K-107. Bucaramanga, Colombia

E-mail: sandra.correa@upb.edu.co

Abstract. This study reports the optimization of innovation chemical techniques in order to improve the remediation of soils contaminated with organochloride pesticides. The techniques used for remediation were dehalogenation and chemical oxidation in soil contaminated by pesticides. These techniques were applied sequentially and combined to evaluate the design optimize the concentration and contact time variables. The soil of this study was collect in cotton crop zone in Agustin Codazzi municipality, Colombia, and its physical properties was measure. The modified dehalogenation technique of EPA was applied on the contaminated soil by adding Sodium Bicarbonate solution at different concentrations and rates during 4, 7 and 14 days, subsequently oxidation technique was implemented by applying a solution of $KMnO_4$ at different concentration and reaction times. Organochlorine were detected by Gas Chromatography analysis coupled Mass Spectrometry and its removals were between 85.4- 90.0 % of compounds such as 4, 4'-DDT, 4,4'-DDD, 4,4-DDE, trans-Clordane y Endrin. These results demonstrate that the technique of dehalogenation with oxidation chemistry can be used for remediation soils contaminated by organochloride pesticides.

1. Introduction

In Colombia, in the municipality of Codazzi, Cesar; still manifest in the environment of organochlorine compounds used in the cultivation of cotton between the 60s and 80s. At the beginning the 90s this agricultural activity declined for various reasons as the El Niño phenomenon, the illicit crops and the economic crisis facing the country at that time [1], these conditions in addition to economic losses, led to pollution of soil resources due to burial of agrochemicals and improper disposal of the remaining chemicals (especially organochlorine pesticides) on adjacent land to the plantations [2]. In addition to the burial of these pesticides a public health problem appear with increasing cases of cancer and congenital diseases. The municipality of Codazzi was considered one of the cities with the highest rate of cancer in the country [3].

Today, much of the affected area is abandoned and neglect by the municipality. It's important to consider that exist studies showing the soil contamination and its effects and has not made any alternative solution to the problem. Therefore, in this study chemical treatments were applied on a laboratory scale, to enable removal of organochlorine in soil. Chemical treatments applied to contaminated soil were based on chemical dehalogenation techniques followed by chemical oxidation. To perform these treatments was applied chemicals in the soil getting results between 80 and 90% removal of pesticides. The soils were characterized by physicochemically and biologically. After that we apply the treatment of dehalogenation and oxidation with soil contact time and concentration of the substance as variables. Analysis of chlorides present in soil were performed using atomic absorption techniques. The analysis to determine the removal of pesticides in soil were performed by gas chromatography coupled to mass spectrometry.

2. Experimental part

Samples of soils contaminated by pesticides were taken at the cotton area located in northern Colombia in the nearby of the city of Agustin Codazzi, Cesar. In this paper we present three stages for remediation of soil contaminated with pesticides. In the first stage physicochemical properties of the soil of interest were identified. In the second stage the techniques of chemical oxidation and dehalogenation were optimized. Finally the removal efficiency of each of these techniques was determined.
The sampling procedure was carried out following the methodology proposed by Corpoica [4]; the affected area was demarcated in two quadrants of 50 x 50 meters with zigzag sampling. The sampling achieved a total of 80 samples collected at 40 cm deep and with a volume of approximately 3 kg per sample. In the laboratory the soil samples were processed by crushing, screening and homogenization to achieve a composite sample.

2.1 Soil physicochemical characterization

For the analysis of the physicochemical characteristics of soil relevant evidence based on standardized methods were performed. The parameters measured were: texture, moisture content, moisture loss, density (apparent and real), total porosity, pH, electrical conductivity and cation exchange capacity (CIC) [5]. The soil was submited to analysis for the detection of organochlorine compounds and their concentrations in the Laboratory of Gas Chromatography and Mass Spectrometry (CROM-MASS) of the Industrial University of Santander (UIS). This analysis was performed using the dispersion technique solid phase matrix (MSPD), using as reference material certified mixture of organochlorine pesticides part No. OCP508-1JM (Chemservice, West Chester, PA 19381 PO BOX 599, USA). The chromatographic analysis was performed on a gas chromatograph AT 6890 Series plus (Techonologies Agilent, Palo Alto, California, USA) coupled to a mass selective detector (Agilent Technologies, MSD 5973) operated in SIM mode [monitoring ions selected]. The column used for analysis was DB-5MS [5% -phenyl-poly (methylsiloxane), 60 m x 0.25 m x 0.25 microns]. The injection was performed in splitless mode ($V_{iny} = 2\ \mu L$).

2.2 Degradation techniques organochlorine

The treatment implement of two chemical techniques applied sequentially on the ground to make the respective decontamination. The techniques used were chemical dehalogenation followed by a chemical oxidation. The optimization of these two techniques was carried out by implementing a 2^2 factorial design (two factors on two levels), which allow to select the best parameters to decontaminate the soil. In both designs the object of study was the soil contaminated with organochlorine pesticides and the response or dependent variable was the concentration of organochlorines.

2.2.1 Chemical Dehalogenation

The technique used was a variation of the chemical dehalogenation by bases catalyzed decomposition developed by the Agency for Environmental Protection (EPA), the method 542-F-96-020 [6]. In the applied modified Sodium Bicarbonate ($NaHCO_3$) solution it was added in different concentrations between 0.5, 1, 5.0 and 10% w / v and further maximum and minimum flow of 0.4 mL/min and 0.10 mL/min respectively in the contaminated soil. The contaminated soil was heated in a temperature range between 38 °C and 40 °C simulating the temperature at which the soil is exposed in their home area. These modifications were made in order that the technique has a bigger applicability and less costs in the area of interest. For dehalogenation process for laboratory scale we used 100 g of contaminated soil in a beaker heated to 38 °C using an electric hot plate, dosing the soil with a solution of sodium bicarbonate ($NaHCO_3$) with concentrations between 0.5 and 10%. The containers with contaminated soil were sealed with parafilm and connected with distilled water in bottles sealed conditions, in order to collect the gases generated by applying the technique. After the dehalogenation chlorides analysis was performed in water by the volumetric method according to Standard Methods numeral 4500-Cl B to confirm the presence of free chlorides from the soil layers as gas and condensed in water. The amount of sodium bicarbonate applied to each test was approximately 70% of field capacity soil.

The 2^2 factorial design allowed us to determine the main effects of the concentration of sodium bicarbonate, the flow rate, and the effect of the interaction of both factors in decreasing concentration of organochlorine pesticides. Each test had three replicas.

The results obtained by the chloride method allows to select the concentrations of sodium bicarbonate and flow rates to be used in contaminated soils. The analyses of removal of contaminants from soil to best conditions selected were submited to the detection of dehalogenated compounds by gas chromatography-mass spectrometry.

2.2.2 Chemical Oxidation

After the implementation of chemical dehalogenation, the soil was treated by chemical oxidation process by applying potassium permanganate as an oxidizing agent [7]. To optimize this technique 2^2 factorial design was implemented and the factors were chosen based on what is reported in literature. The factors to be evaluated were the concentration of the oxidizing agent [$KMNO_4$] and reaction time of soil contaminated with $KMNO_4$. For this technique we used analyzes of samples by gas chromatography.

In the application of oxidation technique on the contaminated soil it consisted of a galvanized metal base with nine carrying vessels of PVC. This material was used for the oxidizing character of potassium permanganate used in the process of soil decontamination using this technique. The dimensions of the vessels of PVC were similar to those of the containers used in the prior art, so that the dehalogenated soil was transferred to these vessels to apply oxidation. PVC cups had an outlet pipe made of the same material for collecting leachate, the output was controlled by a valve and deliver into glass beakers of 100 mL. The aggregate of potassium permanganate for oxidation chemistry correspond to the 30% of the field capacity remaining available in each soil sample. Thus the soil was 100% field capacity, tests were also performed in triplicate. The factorial structure 2^2 applied to the oxidation of organochlorine compounds allow to identify the main effects of the concentration of the oxidizing agent ($KMnO_4$) between 1, 0.5, 0.25 and 0.01% p/v, the reaction time, and the oxidant interaction effect of both factors in decreasing concentrations of organochlorines. The samples with different concentrations of $KMnO_4$ were evaluated 7 days after the application of the oxidizer to the ground in accordance with the literature reccomendation. Once determined the concentrations of potassium permanganate we proceeded to determine the reaction time for the optimization of oxidation were between 1, 0.5, 0.25 and 0.01% p/v. The results of gas chromatography analysis allows to choose the best combination for chemical oxidation.

2.3 Determination of the efficiency of removal of organochlorine

The gas chromatography analysis allows to determine the efficiency of removal of organochlorine compounds to each of the techniques implemented. In chemical dehalogenation, they were selected factorial design variables by determining chlorides in soil samples latter were submited to gas chromatography analysis to establish the efficiency of removal of each variable. Similarly, the removal efficiency of chemical oxidation was calculated from the results obtained by gas chromatography. To calculate the removal techniques we consider the baseline concentrations of pesticides, determined by the soil chromatographic analysis.

3. Results and discussion

3.1 Physical and chemical Soil characterization

The physicochemical analysis of the soil favourably loamy described for vegetation [8], with a neutral pH optimum of 6.68 for the solubility of essential nutrients for plants and electrical conductivity irrelevant, however signs of degradation evidenced by compacting the value the Da (1.604 g / cm3) and total porosity (23.45%), along with a low content of organic matter (1.08%) Average cation exchange capacity (17.6 meq / 100 g soil) nutrient deficiency and (P and N).

Tabla 1. Physical and chemical properties in soil contaminated

PROPERTY	PARAMETER	RESULT	INTERPRETATION
PHYSICAL	Texture	Well graded Franco	Intermediate particle size and wide variation.
	Humidity (%)	HG: 2.43; HV: 3.9	low moisture
	Density (g/cm^3)	Dr: 1.93; Da: 1.604	Degradation by light compaction to medium
	Porosity (%)	23.45	
	Field capacity (mLH_2O/KgS)	227.7	Medium liquid retention
CHEMICAL	pH (U)	H_2O: 6.68; KCL: 6.26; ΔpH: -0.42	Neutral negative colloid charge and cation exchange
	Conductivity (µs/cm)	204.6	Negligible in altering plant growth.
	CIC (meqg/100gS)	17.6	CIC Media, limiting availability of essential cations
	CO y MO (%)	CO: 1.06 MO: 1.802	Low content of organic component
	Nutrients (mg/Kg)	N total: 1025.55 P total: 1.33	Poor content of nutrients

According to the results of the soil is classified as low or mineral ore with low content of organic matter. Notably, the climate of the area where the soil is important for the percentage of organic matter, for warm climates or where there are high temperatures, organic matter is degraded more rapidly due to microbial activity present [9]. Moreover, low in organic matter in the soil as associated with the limited capacity of cationic exchange of the soil. The soil shows poor fertility therefore low organic matter, signs of degradation by compaction. Table 2 organochlorine pesticides found in the soil sample under study and their concentrations, obtained by analyzing chromatography coupled with mass spectrometry gases occurs.

Table 2. Initial concentrations of pesticides in soil of interest

Compound	Concentration of pesticide in sample (mg/kg)
trans-Clordano	0.1
4,4´-DDE	1.1
Endrin	1.9
4,4´-DDD	2.4
4,4´-DDT	9.1

The results show the presence of concentrations of the pesticide DDT and its metabolites DDD and DDE degradation, and other pesticides such as trans-chlordane and endrin. The present study has a primary focus on DDT to be the most toxic compound and its two metabolites DDD and DDE degradation.

3.2 Optimization degradation techniques

The results show the presence of concentrations of the pesticide DDT and its Analyzed variables in the design of experiments established that the chemical dehalogenation of pollutants optimal concentration of sodium bicarbonate and a dosing flow were optimized for the technique. For oxidation evaluating the concentration of potassium permanganate and the contact time with this reagent soil determined the best conditions for chemical oxidation ground.

3.2.1 Chemical deshalogenation in soil

In Figure 1, a progressive increase in chloride concentrations in soil with increasing concentration of sodium bicarbonate is recorded, indicating that there is a correlation between the two parameters. In the analyzed results with high variability values are recorded, this may be because the Cl- ions in the soil were released by the dehalogenation of organochlorine compounds in the course of time and may be associated Cl_2.

Removal concentration of chlorides in soil was obtained at different concentrations of sodium bicarbonate solution in range of 0.5% w/v (lowest) to 10% w/v (highest). Additional different flow rates corresponding to 0.4 ml/min (Fmin) and 10 ml/min (Fmax) was applied. Analysis of chlorides in soil under these conditions was presented in Figure 1.

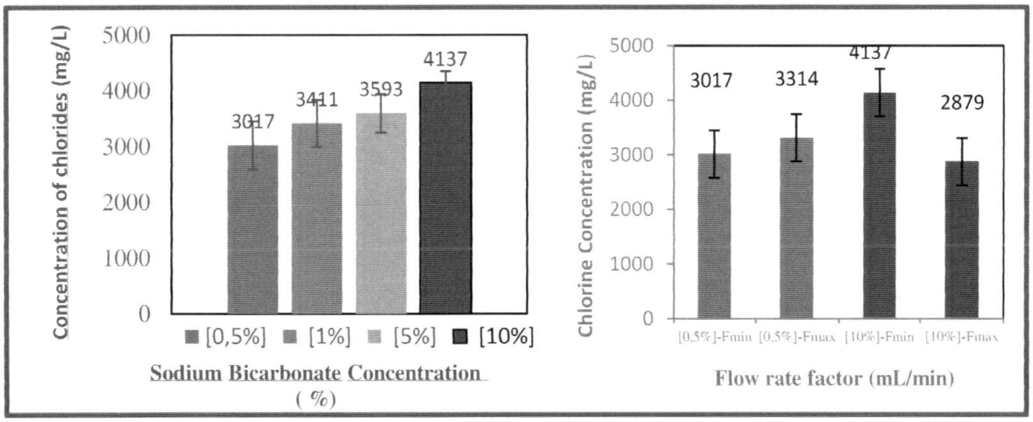

Figure 1. Chemical deshalogenation treatment according sodium bicarbonate concentration and the flow rate factor.

The presence of chlorides after application dehalogenation confirms breaking the molecule organochlorine and release of halogen (Cl), reducing the toxicity of pesticides in soil. The results for the initial concentrations of trans-Chlordane, 4,4'-DDE, Endrin, 4,4'-DDD and 4,4'-DDT demonstrated the decrease evident in the compounds concentrations in all samples analyzed. The values obtained by applying the bicarbonate solution 0.5% w/v were the most significant in both high and low flow with minimal difference in concentrations suggesting that the flow rate factor is not representative. However the combination of sodium bicarbonate $NaHCO_3$ [0.5%]-Fmax, presented better results in the dehalogenation of all organochlorine compounds evaluated.

The 4,4'-DDT in the effect of the concentration of sodium bicarbonate ($NaHCO_3$) and the effect of flow rate are positive, suggesting that increasing the concentration of sodium bicarbonate and the flow rate will increase the concentration of this pollutant, also the effect of the interaction of both factors is positive suggesting same effect generation of increased concentration of 4,4'-DDT in soil. On the other hand it is important to emphasize that the results greater magnitude in the most significant effect is the concentration of $NaHCO_3$ used.

To summarize the results of the application of design 2^2 for the three organochlorine compounds of interest suggest that the factor concentration of sodium bicarbonate ($NaHCO_3$), is the most important in the removal of these compounds at lower concentrations their decline in soil. After applied dehalogenation were proceeded to testing chemical oxidation technique. For determination of this factor four trials using different concentrations of $KMnO_4$ (1, 0.5, 0.25, 0.01% w / v) and analyzed after seven days contact between the oxidant [10] and the contaminated soil is made, the analysis of they performed by gas chromatography. Table 3 shows the contaminant concentration occurs den soil after applying a removal with potassium permanganate oxidant at different concentrations.

Table 3. Chemical oxidation treatment applied to contaminated soil by pesticides

Compound	Concentration of Potassium Permanganate			
	[1%]	[0,5%]	[0.25%]	[0.01%]
trans-Clordano	0.09	0.05	0.05	0.05
4,4´-DDE	0.65	1.3	0.74	0.61
Endrin	0.64	2.78	0.74	0.68
4,4´-DDD	8.45	1.99	0.96	0.84
4,4´-DDT	1.46	3.23	1.18	1.08

According to the results of Table 3 oxidation with $KMnO_4$ to 0.25 and 0.01% w/v shows that the concentrations of the compounds had a uniform descending behavior, whereas oxidation with $KMnO_4$ and 0.05 to 1% w/v report outliers where concentrations of contaminants up again or rebound. According to the EPA this occurs when oxidants injected had no contact with any contaminated area or the oxidant is exhausted before trying all evidenced based contamination. In concentrations chosen for the oxidation process were 0.25 and 0.01% w/v.

The concentrations selected of potassium permanganate to 0.25 and 0.01% w/v were evaluated at three times 4, 7 and 14 days. Table 4 present the pesticide concentrations after applying oxidation with $KMnO_4$ in soil with pretreatment of dehalogenation.

Table 4. Concentrations of pesticides in soil by oxidation treatment using 0.25 and 0.01% $KMnO_4$.

Compound	**Chemical Oxidation by $KMnO_4$ 0.25%**			**Chemical Oxidation by $KMnO_4$ 0.25%**		
	4 days	**7 days**	**14 days**	**4 days**	**7 days**	**14 days**
trans-Clordano	0.02	0.08	0.06	0.06	0.08	0.05
4,4´-DDE	0.54	1.05	0.93	0.74	1.08	0.59
Endrin	0.6	0.93	0.72	0.52	0.74	0.48
4,4´-DDD	0.82	1.41	0.75	1.16	1.52	0.84
4,4´-DDT	1.82	3.08	2.82	1.33	2.99	1.61

The results of chemical oxidation with potassium permanganate show inconsistency in the behavior, for this reason was chosen as the criterion for selecting the combination of factors for the chemical oxidation which showed more favorable results for the elimination of the compound with increased toxicity , 4,4'-DDT. Table 4 shows that the sample with $KMnO_4$ [0.01%] analyzed after 4 days reports the best result for the 4,4'-DDT, however does not generate the same effect for 4,4'-DDD and 4 , 4'-DDE.

As shown in Table 4, the samples in a contact time of 7 days with concentration of sodium permanganate and 0.25 to 0.01% w/v increasing concentrations recorded for the compounds tested and significantly with respect DDT a contact time of 4 days and 14 days. The results demonstrating the phenomenon of increased concentrations of contaminants referenced by the EPA, which in turn is disadvantageous for the removal of compounds.

3.3 *Efficiency of removal of organochlorine*

Based on the concentrations detected by gas chromatographic analysis of each applied techniques and the initial concentrations of pesticides in soil, the trend removal of organochlorine interest soil was analyzed. Figure 2 shows the trend of reduction the concentrations of organochlorine compounds by comparison the techniques applied previously selected conditions.

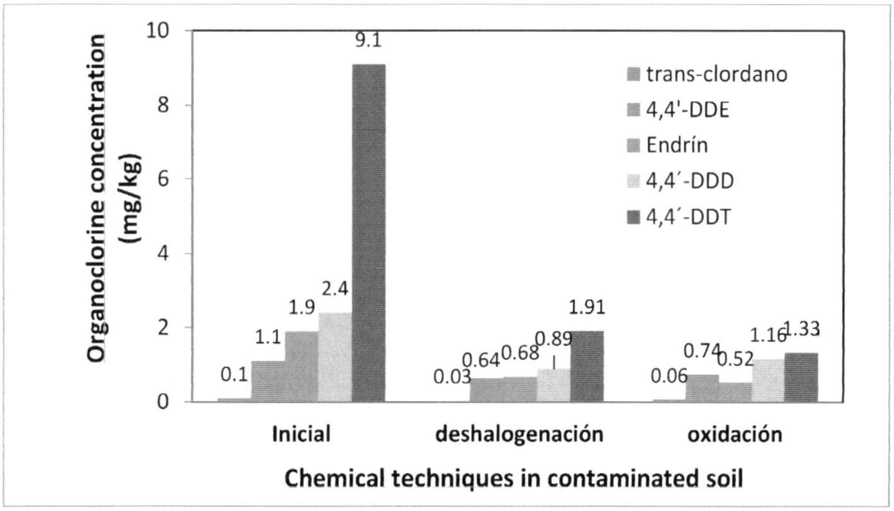

Figure 2. Concentration of organochlorine pesticides during chemical treatment phases.

In comparison by other techniques in specific biology treatments, the deshalogenation treatment applied in soil contaminated by pesticides is more effective in short times from 4 to 7 days. Biological techniques achieve the same removal percentages but in times of 3 to 9 months. However the chemical techniques can cause loss of soil´s microorganisms and require the application of additional techniques. The application of chemical oxidation is more effective than dehalogenation technique, but obligate to move exsitu the soil, which implies increasing cost and additional treatments.

4. Conclusions

The physicochemical analysis of the soil described for greening loam with a neutral pH optimum of 6.68 for the solubility of essential nutrients for plants and electrical conductivity irrelevant. However signs of degradation by compaction highlighted by the value of the Da (1.604 g/cm^3) and total porosity (23.45%), along with a low content of organic matter (1.08%), average cation exchange capacity (17.6 meqg / 100 g soil) and deficiency of nutrients (P and N).

Deshalogenation technique proposed has favorable results for the lower concentrations of o organochlorine in the ground reaching up to 81.8%. The analysis suggested that the analysis 2^2 factor sodium bicarbonate concentration is more significant than the flow rate used, revealing that the lower the level of the chemical concentration greater reduction was obtained in all organochlorine.

In chemical oxidation with $KMnO_4$ increases, it is evident in the organochlorine concentrations in some samples, attributing this phenomenon to incomplete distribution of the oxidizing agent in the matrix of the contaminated soil. In the 2^2 factorial design it was determined inconsistency in the behavior of this technique, however oxidation can remove the 4,4'-DDT, DDE 4,4'- and 4,4'-DDD to different conditions each.

The removal percentages obtained in the first phase of treatment comprised dehalogenation technique were 79.0% for 4,4'-DDT, 41.8% for 4,4'-DDE and 62.9% for 4,4'-DDD. The second phase comprised by chemical oxidation favored continuous reduction of the compound 4,4'-DDT reaching up to 85.4% removal, but generated a decrease of the previously obtained removals of 4,4'- and 4,4'-DDE y DDD to values of 32.7% and 51.7% respectively.

Acknowledgments

We would like to thank the Pontifical Bolivarian University of Bucaramanga sectional for the financial support of this project through funding for the improvement of science (DGI-BI053B project) and Gas Chromatographic Lab Industrial University.

5. Reference

[1] Solano S P 2009 *Cuadernos Des. Rural*, 6 (62): 53-79
[2] Sánchez N P et al 2006 *Revista de Ingeniería Universidad de los Andes*. **23**. 13
[3] Al-Rawas A A and Goosen M 2006 *Expansive Soils. Recent advances in characterization and treatment* (London) vol 1 (UK: Taylor & Francis)
[4] USEPA SW846 *Guideline on Laboratory* Analysis of Potentially Contaminated Soils, Sampling and analysis of waters, wastewaters, soils and wastes.
[5] Salazar J C 2000 *Soil Quality Test Kit Guide* USDA Departamento de Agricultura de los Estados Unidos.
[6] US EPA (United States Environmental Protection Agency) *Deshalogenación Química* EPA 542-F-01-010S 2002
[7] Chiron S and Fernandez-Alba 2000 *Water Research*. **34** 366
[8] Yong R and Mullingan C N 2004 *Natural Attenuation of contaminants in Soils* (Lewis Publishers)
[9] Pierzynski G M 2005 *Soils and Environmental Quality* (Taylor & Francis)
[10] Siegrist R L et al 2011 *In situ Chemical Oxidation for Groundwater Remediation* (New York) vol 1 (USA: Springer)

Design, construction and evaluation of a system of forced solar water heating.

E Hernández[1], G A Bautista[1], I L Ortiz[1]

[1]Universidad Pontificia Bolivariana, Bucaramanga, Colombia

E-mail:emil.hernandez@upb.edu.co

Abstract. The main purpose of this project was to design, construct and evaluate a system of forced solar water heating for domestic consumption, at the Universidad Pontificia Bolivariana-Bucaramanga, Colombia; using solar energy. This is a totally system independent of the electrical grid and an important characteristic is the heating water doesn't mix with the consumption water. The system receives the solar radiation through a flat-plate collector, which it transmits the heat to the water that it flow with impulse from the centrifugal pump of 12VDC, the water circulates toward helical serpentine it is inside of the tank of the storage whose capacity is 100 liters of water. The temperature of the tank is regulated with a controller in such a way that de-energized the pump when it gets the temperature required. The performance thermal or efficiency of the system was evaluated like a relationship between the delivered energy to the water in storage tank and the incident energy in the flat-plate collector.

1. Introduction

In the publication by Chuck Marken, 2009 [1], a foundation on solar energy exchanger is made, which transfer heat from one fluid to another without contact between them. The most important design parameters are the wall material should be a good conductor, the contact area, the physical configuration of coil or flat pipe, the flow rate, and the type of fluid, generally water depending on the weather will additives added. Brian Mehalic [2], speaks of the need to install pressure relief valves in the tank, because the water temperature increases and pressure causing a possible rupture of the tank, therefore keep a good coating for long periods of time the temperature in the tank and the material must be resistant to maintain suitable conditions, water consumption is required. In 2007 an American professor builds a forced flow heater (system that was implemented in this project) and James Dontje [3] describes step by step each of the determinations that led to the construction of this system. The first step was the choice of system, he ruled out direct systems which will expose a pipe to freeze.

The thermosiphon system was out of reach for the need to place the tank below the collector, he thought of a photovoltaic system powered by a DC pump, closed loop which used distilled water and create drains pipe to prevent damage when the water freezes . One of the big challenges was to reach the desired pump head, as already being in the maximum height, beat other resistance is relatively easy. There is little head is expected to pump current and usually more striking pumps AC found in greater quantity, variety and low cost, but the ultimate goal was the independence of the network, so it had to work to find the DC pump indicated. Prior to mounting the pump, the required maximum height was measured and was a test to see if the pump was able to reach it. Indeed he did, including the bypass, which served to beat other resistance. The rest was working plumbing, electrical installations and protection. Getting to the end of the first day at 90 gallons of water at 50 ° C.

Numerous studies have investigated the performance of solar water heaters operating by thermosiphon [4-7]. Burbano et al. [4] they built a thermosyphon solar water heater with an interesting design process. Zerrouki A. et al. [5] found experimentally that the rate of mass flow increases in the hour of greatest radiation (11 am-2pm). Sakhrieh and Al-Ghandoor [6] presented an experimental study of the performance of different types of solar collectors and conclude that the vacuum tube collector is the top performer with respect to efficiency, with a maximum value of 72%. Gupta and Garg [7] experimentally proved that the flow rate of a thermosyphon water heater can be increased if the relative height between the collector and the storage tank is increased, but the efficiency is not increased.

2. Theory

2.1 Energetic balance of a flat collector

$$Q_1 = Q + Q_2 \tag{1}$$

Q_1: is the incident energy (direct + diffuse + albedo) in the unit time.
Q: is useful energy, by collecting the heat transfer fluid.
Q_2: is the energy lost by dissipation to the environment.
Given the definition of intensity of radiant energy, Q_1 is the product of the intensity and the surface. Not all the incident energy Q_1 will be absorbed in the absorber. First, if there is deck, you have to have the transmittance in a count, which will leave to flow part of this energy (τSI). Moreover, the absorption coefficient α or absorptance of the absorber plate never becomes equal to unity, so that the fraction of energy actually absorbed is:

$$Q_1 = \tau\alpha SI \tag{2}$$

As for the energy lost Q_2, detailed calculation is very complex because, must be taken into account simultaneously and in different proportions losses by radiation, convection and conduction. However, in order to be able to use a simple formulation, it has been agreed to include these influences in the overall loss coefficient called U, which is measured experimentally and is data supplied by the manufacturer. Experience has shown that supposing losses per unit of surface proportional to the difference between the average temperatures t_C of the absorber plate and the environment t_a, being the proportionality factor the coefficient U, is a good approximation. So that:

$$Q_2 = SU\,(t_c - t_a) \tag{3}$$

Substituting (2) and (3) in (1) is obtained:

$$Q = S\,[I(\tau\,\alpha) - U(t_c - t_a)] \tag{4}$$

S: collector surface
I: radiant intensity [W / m^2]
τ: transmittance of the transparent cover
α: absorptance of the absorber plate
U: overall loss coefficient [W / (m^2 ° C)]
t_c : average temperature of the absorbing plate [° C]

t_a: environment temperature [° C]
The efficiency ratio of the collector, η, is defined by the relationship between energy captured and received at a given instant according to the equation (5).

$$\eta = Q / [SI] \tag{5}$$

Substituting (4) in (5) is obtained (6)

$$\eta = F_R[(\tau\alpha)-U(t_m-t_a)/I] \tag{6}$$

Where t_m is the mean temperature of the collector according to the equation (7).

$$t_m = (t_e + t_s) /2 \tag{7}$$

For a flat plate collector, the equation (6) is approximate to the equation (8) [8].

$$\eta= 0.83 - 6.8(t_m - t_a) / I \tag{8}$$

2.2 Efficiency of the system
η= Delivered Energy /Received Energy from collector
$$\eta= [qt / t] / [Ac * Gt] \tag{9}$$
qt : Quantity of received heat by stored water
t : Time where hot is received (was taken 1 h)
Ac : collector Area
Gt : Average radiation during the time that tank received radiation
qt = m*Cp* (Tf – To)
Tf: End Temperature of water in the tank.
To: Initial Temperature of water in the tank.

3. Methodology

The system built solar heating forced shown in Figure 1. The solar collector has the dimensions 2.0 m * 1.2 m * 0.13 m with stainless steel plates and copper according to Figure 1. Internally the collector has twelve copper tubes. The collector has a tempered glass dimensions 1.6 m * 0.8 m, which determines the catchment area of solar radiation.

The water heating system shown in Figure 1, is performed in a cycle that involves two stages, in the first water from the external tank it flows through the pump to the flat plate collector and take heat from it in the second stage will he sent water to a helical coil that is inside the storage tank, there while the water flows through the coil transmits heat acquired in the collector to the drinking water, finally the water is discharged into the external tank. This process is carried out in a loop driven by a 12V DC centrifugal pump and de-energizes the pump controlled by a pyrometer that when the temperature is ideal. Table 1 shows the results of calculating the efficiency of the system of the Figure 1.

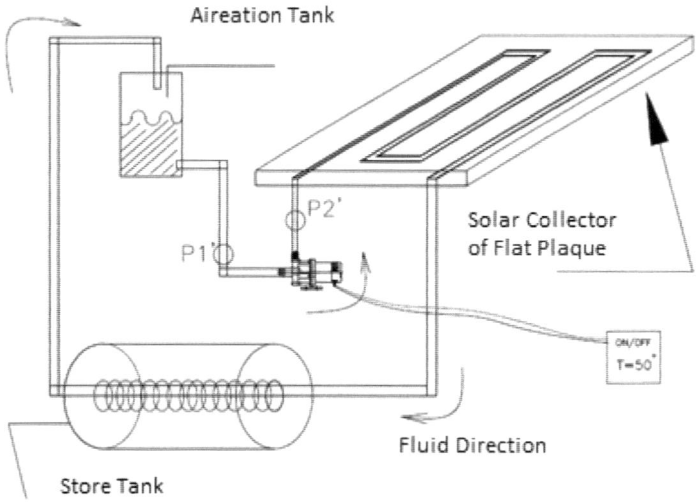

Figure 1. Implemented forced solar water heating system

Table 1. System Efficiency

Day	To [°C]	Tf [°C]	Ac [m^2]	$\overline{Gt}$ [W/m^2]	ΔT	qt [J]	*Efficiency* [%]
24	32,2	35,2	2,4	**786**	3	368,2	19,5
25	40,8	44,5	2,4	874	3,7	454	21,6
26	44,9	47,8	2,4	1877	2,9	354	18,9
27	46,4	49,5	2,4	1097	3,1	380	14,4
28	49,3	55	2,4	1125	5,7	700	25,9
31	37,4	39,4	2,4	574	1,9	233	16,9

Efficiency in Table 1, is the best obtained efficiency that day, calculated according the equation (9) and the experience was made from 24 to 31 August of 2015.

Table 2. Collector Efficiency

Te [°C]	Ts [°C]	Ta [°C]	I [W/m^2]	*Efficiency* [%]
38	39	28	**618**	71.0%
42	44	28.5	624	67.0%
45	48	27	624	61.7%
47	49	29	652	63.0%
50	51	28	650	59.0%
50	52	26	648	56.0%

Efficiency in Table 2, was obtained every 15 minutes and is showed because was obtained the best efficiency of 71.0%, calculated according the equation (8).

4. Conclusions

The system of forced solar water heating was implemented, which provides hot water for domestic use with a capacity of 106L of water, raising the temperature from room temperature to 55 ° C. The maximum efficiency for the system was 25.9 % and for the collector was 71 % this are good results because the accepted commercially efficiencies are approximated [8-10]. The storage tank maintains the isolation because the left over a time of 18 hours at environment temperature, water temperature decreased 4 ° C.

References

[1] Home Power Magazine. Solar heat exchanger equipment. Available at: http://www.homepower.com/articles/solar-water-heating/equipment-products/fundamentals-solar-heat-exchangers. (Accessed 15 August 2015).

[2] Home Power Magazine. Solar-hot- water equipment. Available at: http://www.homepower.com/articles/solar- water-heating/equipment-products/solar-hot-water-storage. (Accessed 8 September 2015).

[3] Home Power Magazine. Solar hot water system. Available at: http://www.homepower.com/articles/solar- water-heating/project-profiles/designing-pv-powered-drainback-solar-hot-water-system. (Accessed 30 August 2015).

[4] C Burbano, H Restrepo, y J Sabogal 2006 *Sciencia et Technica,* **31,** p 85.

[6] A Sakhrieh, A Al-Ghandoor 2013 *Energy Conversion and Management*, **65,** p 715.

[7] Cl Gupta, HP Garg 1968 *Solar Energy,* **12,** p 163.

[8] H Amaya, L Solano, 2014 *Diseño, construcción y evaluación de un sistema de calentamiento de agua, con un arreglo en serie de colectores de placa plana, utilizando energía solar*, p 79.

[9] D Arias, E Villadiego, 2015 *Comparación de eficiencias de tres sistemas de calentamiento solar,* p 54.

[10] J Burbano, A Restrepo, 2006 *Diseño y construcción de un calentador solar de agua operando por termosifón,* p 66

AUTHOR INDEX